DATE DUE			
Sep21'76			
Apr19 '82			

ENTERING INDUSTRY

ENTERING INDUSTRY

A GUIDE FOR YOUNG PROFESSIONALS

FRED W. BILLMEYER, JR.

Professor of Chemistry
Rensselaer Polytechnic Institute
Troy, New York

and

RICHARD N. KELLEY

Kodak Park Division
Eastman Kodak Company
Rochester, New York

A Wiley-Interscience Publication

JOHN WILEY & SONS

New York / London / Sydney / Toronto

Library of Congress Cataloging in Publication Data:

Billmeyer, Fred W.
Entering industry.

"A Wiley-Interscience publication."
Includes index.
1. Engineering as a profession. 2. Chemical engineering as a profession. I. Kelley, Richard N., 1940- joint author. II. Title.
TA157.B5 620′.0023 75-22283
ISBN 0-471-07285-0

Printed in the United States of America

10 9 8 7 6 5 4 3 2 1

PREFACE

This book is designed to guide you, young professionals, in developing your acquaintance with industry, where some four-fifths of you will spend your working careers. We deliberately use the words "young professional" to describe you:

If you have or are earning a college degree in the sciences or engineering, at any level, then you are a *professional* by any accepted definition of the word. And you should be proud of it.

If you are still in the university, at any level from undergraduate through graduate, postdoctoral, and young staff member, to the senior faculty member with a well-developed sense of responsibility to the students; if you have recently entered industry but haven't yet mastered all its ins and outs, learned all you ought to know about it, found your own place in it to your full and lasting satisfaction; or if you have done all these things but are still young in heart, retaining that important desire to find out what makes things work and how you fit into the picture, then you are *young* by our definition.

What does *Entering Industry* contain? We could just ask you to turn the page to the table of contents, but we hope you won't quite yet because there are a few ground rules, also mentioned in the text, that we'd like to emphasize.

Entering Industry generally describes the types of companies, large and small, that employ professionals in science and engineering to aid in the manufacture of products or the provision of services. Our approach is based on the premise that you are contemplating entering industry, or entered it recently. We start with a brief description of the industry we have selected as an example, talk next about how you should go about getting a job or changing jobs, then about your responsibilities as a young industrial professional, and your routes and opportunities for advancement. In the remaining chapters—about half of the book—we take the typical industrial corporation apart and analyze it, division by division, in approximate order of decreasing interest and immediate job opportunities for the scientist or engineer. We begin with research and development, progress through manufacturing, marketing, and the other divisions (with a side excursion for a close look at patents), and culminate in management.

Our objective throughout is to supply you with the information you will need to direct your education toward an industrial career, find employment in the company of your choice, and advance in a satisfying life as a professional in industry.

For convenience and continuity, and because it is familiar to us, we have selected the American chemical industry as an example on which to base *Entering Industry*. Our education and industrial experience has been in that industry, and it would have been foolhardy for us to select another, or to skip from one to another, whenever we wished to provide an illustration of a point.

We hope this doesn't turn you off if your professional education is in another field of science or engineering. On the contrary, we encourage you to look on this as a bit of a challenge. We have, we think, been fair in dealing with the chemical industry. Using equivalent sources of information about your own field, many of which can be located from clues in this book, we suggest that you form your own assessment of your discipline; search out and pinpoint its strong and weak points; and plan your career accordingly.

Throughout *Entering Industry* we have stressed the importance of your ability to communicate, and as best we could, provided examples in our own writing. But the English language has some limitations, one of which is the absence of a single word encompassing both he and she, his and hers, and so on. In this day when professional women should be fully equal to their male counterparts, we apologize for using only the male pronouns where inserting both repeatedly would lead to awkwardness. It is not male chauvinism showing!

We also believe in the old pedagogical custom of telling you what we are going to tell you, then telling you, then telling you what we've told you. So you will find some deliberate repetition in this book; bear with it, for it points out the facts and opinions we wish to stress.

We have expressed our own views and conclusions in this book; they are not necessarily those of our employers or any other organization with which we have been or may be affiliated. We have tried to indicate clearly where opinions are our own, and to provide references where they are not. We feel rather strongly about providing an adequate guide to the literature; nothing annoys us more than a book full of unsubstantiated facts, with no clues as to where to go to obtain more information. This we have tried to avoid.

We wish to thank our colleagues and associates, at Kodak, Rensselaer, and elsewhere, for helpful discussions and comments; in particular we are grateful to John M. Calhoun, Hans Coll, James W. Geriak, Fred W. Hoyt, Emil W. Milan, and Walter R. White. Special thanks go to Claudia L. LeBarron for typing the manuscript and to Vivian F. Capo and Mary T. Lincoln. This book could never have been written without the patience and understanding of our wives, Annette and Felicia, and our children, who have endured our absence and preoccupation many more times than we would have wished.

FRED W. BILLMEYER, JR.
RICHARD N. KELLEY

Troy, New York
Rochester, New York
June 1975

CONTENTS

ENTERING INDUSTRY

CHAPTER 1

INTRODUCTION

It's likely that many of you young professionals picking up this book are nearing the end of the formal educational phase of your life and beginning to look for a job. Some of you, already employed, may be thinking about a possible change. And perhaps still others of you are satisfied, but wonder what the other guy's lot is like.

So you pick up the weekly trade journal and read:

> Industry recruitment of college graduates was up strongly this year, with chemical and drug firms topping all other industry groups in their number of offers to bachelor degree candidates. . . . Job offers to doctoral-level chemists were 45% higher and for doctoral-level chemical engineers 70% higher than a year ago (1).

Or, maybe just now things aren't quite so rosy:

> Chemical employment indexes have declined this year . . . some chemists and engineers have fallen victims to cutbacks. . . . The net result is that . . . new jobs will be much harder to find—for college graduates and seasoned career professionals alike (2).

Regardless of the ups and downs of the job market, there is one fact that comes through every time: The odds are about two to one that as a young professional you will end up in industry, rather than a government or academic job, at the end of your educational career.

So, let's have a look at that industry, see how it operates, and try to find the answers to questions like these:

- What kind of training should I have in school to prepare for an industrial job?
- How do I land a job? (It *won't* be handed to you on a silver platter, *that* we'll say now.)
- What will I actually be doing?

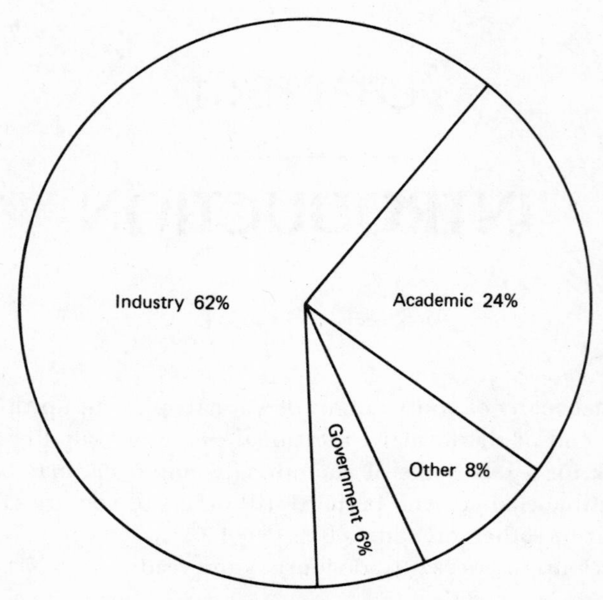

FIGURE 1. Employment of chemists, 1972—you have almost a 2:1 chance of ending up in industry (3).

- How will my performance be evaluated?
- How about that "employment contract" they want me to sign?

Before we dig in, let's set the stage a bit more. Figure 1 justifies our two-to-one odds in favor of your ending up in industry; these figures are for 1972, but they don't change appreciably from year to year. In Fig. 2 we see that within industry there is roughly an even split among positions in research and development (R&D), management, and everything else. (Here "everything else" means all sorts of support groups outside of R&D, ranging from the analytical labs in the manufacturing division, through purchasing, patents, and personnel, to marketing.) Most scientists start their career in R&D, because of its close relation to their academic work. The first position may well be a stepping stone into another sort of industrial job, or into management, after a few years.

We assume that you have already made a professional commitment to a career in science or engineering. Despite all the talk about plumbers and truck drivers, you won't be doing too badly in salary compared to other professionals (Fig. 3), and you will find (Fig. 4) that graduates in only a few mathematically oriented fields among the sciences command higher salaries than industrial chemists and engineers.

So, hopefully, you are on your way to a good, intellectually challeng-

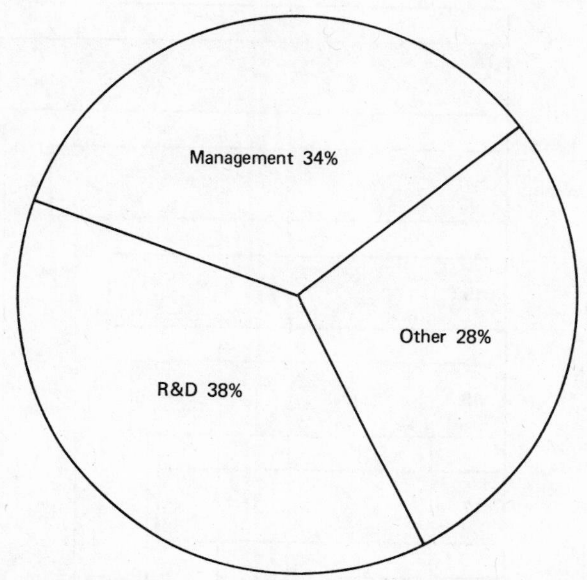

FIGURE 2. Chemists' employment within industry, also 1972—almost an even split among R&D, management, and everything else (3).

ing, well-paid life. And we want to give you a little preview of what it should be like.

The American chemical industry goes back to Pilgrim times, when everyday needs required local sources of a few chemicals—saltpeter for gunpowder, alum for tanning, and potash for soap. By the early nineteenth century, patent protection had been introduced and easier transportation and the western movement were changing the picture from a local to a national one. The product mix expanded too: Early key products were sulfuric acid, white lead for paints, and lye from soda ash. Chemical fertilizers were introduced around 1850, vulcanized natural rubber soon after, and then major processes for soda, chlorine, and ammonia. World War I forced a tremendous expansion on the industry, which had to provide nitric acid for guncotton, and dyes, drugs, and other organic chemicals to replace those previously imported from Germany.

About this time laboratory research was introduced into industry on a large scale, and the payoffs began rolling out: plastics, synthetic fibers, detergents, adhesives; all sorts of petrochemicals, pesticides, and pharmaceuticals; and, under the pressure of World War II, antibiotics, huge amounts of synthetic rubbers, and many others. Today the industry's

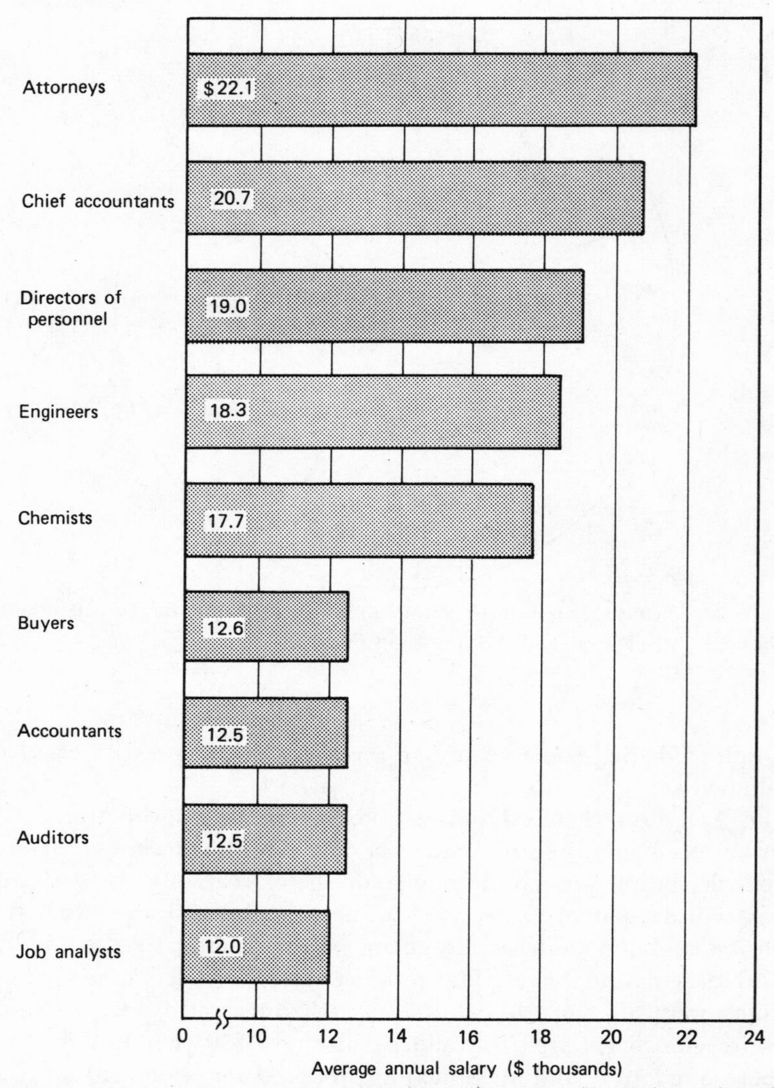

FIGURE 3. Rank of chemists and engineers among salaries of selected white-collar groups, 1973—you stand to be considerably above the average (4). (Reprinted by permission of *Chemical & Engineering News*.)

efforts to turn out exotic new products must be tempered by the pressures of the energy crunch and increased government regulation, as well as humanistic and ecological concerns.

And what about the role of the university with respect to the chemical industry? You are presumably in the one (or left it not too long ago),

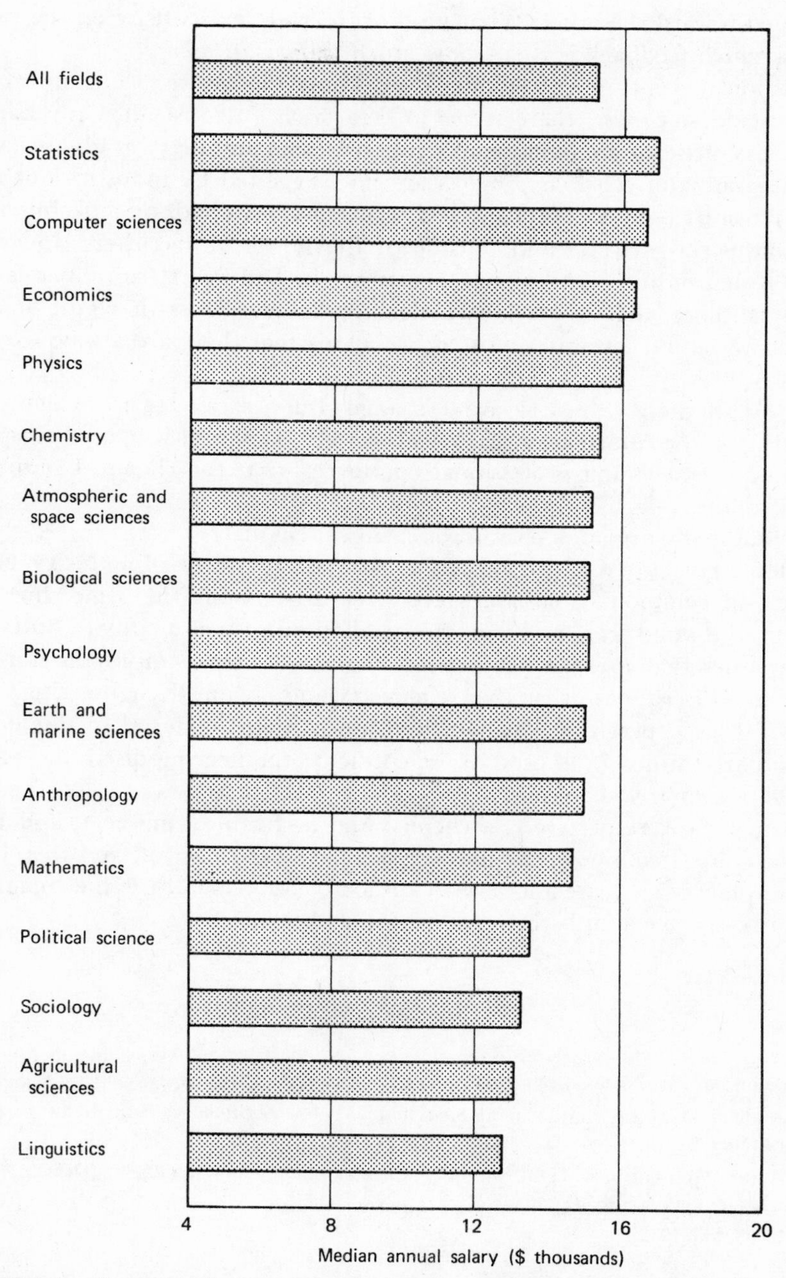

FIGURE 4. Chemists' salaries compared to those of other scientists, 1972—again, well toward the top (3). (Reprinted by permission of *Chemical & Engineering News*.)

headed toward the other. What kind of interface exists between the two? How much feedback is there, how much interrelation?

If you've kept your eyes open, you know the answer: The interface almost doesn't exist; there seems to be a great gulf instead. Universities get very little of their monetary support from industry; graduate (and undergraduate) research projects are not suggested by industry, nor are their results used by industry; few professors have had, or look forward to, industrial careers; and outside of the horde of recruiters (in good years) descending like hungry locusts or (in bad years) conspicuous by their absence, there's precious little interchange across that gulf at all. (Lest we be too severely censured, we admit that there are always exceptions.)

Most of these things we can't change. But we can try to bridge one small gap, the information gap. And so we'll try to tell you, as a prospective or recent young professional employee, what the chemical industry is all about.

What if you are not a chemist of chemical engineer?

Don't give up yet. We recognize that all branches of industry have much in common. Companies are structured about the same, hire in about the same way, and operate in about the same manner. And the companies that make up the chemical industry have employees trained in many disciplines other than chemistry and chemical engineering.

Thus we expect that you will find much useful material in this book if you are trained in almost any scientific or engineering discipline—and perhaps even if you are not.

But we are, respectively, a chemist and a chemical engineer, and this is the subject we know, the industry we have been part of, and therefore the topic we can write about with authority. So we shall use the chemical industry as an example throughout this book.

REFERENCES

1. Anon., *Chem. & Eng. News* **52** (31), 6 (Aug. 5, 1974).
2. Anon., "Chemical Employment Situation Tightens," *Chem. & Eng. News* **47** (49), 22 (Nov. 24, 1969).
3. Anon., "Job Outlook Tight for 1973," *Chem. & Eng. News* **50** (40), 10 (Oct. 2, 1972).
4. Anon., "1973 Salaries Generally Up but Some Down," *Chem. & Eng. News* **51** (41), 13 (Oct. 8, 1973).

CHAPTER 2

THE AMERICAN CHEMICAL INDUSTRY

If we are going to help you young professionals size up the American chemical industry, and find your slot in it, we have to tell you what it is. This is surprisingly difficult.

Everybody has his own idea of what to include in the makeup of the chemical industry. Since we are primarily concerned with your employment in that industry, our objective is to include in our definition any kind of a company that employs appreciable numbers of chemists or chemical engineers. But that definition encompasses a very broad group of categories indeed.

The U.S. Department of Commerce, which compiles and issues many facts and figures about the industry, uses two categories for the chemical industry. The narrower is *Industrial Chemicals,* which includes only such products as alkalis, chlorine, ammonia, nitric acid and their derivatives, basic inorganic and organic compounds, industrial gases, dyes and pigments, flavors and fragrances, and a few others. The Federal Reserve Board's definition of *Industrial Chemicals* adds to these polymers, including plastics and synthetic fibers and elastomers. The Department of Commerce's second and broader category is *Chemicals and Allied Products,* including all of the products above plus a wide variety of less basic or formulated chemical products, such as bulk medicinals, cleaning compounds, fertilizers, and adhesives; and products such as cosmetics, paints, and printing inks.

Yet these categories, the figures for which we will quote because they are about all that are readily available, are still not broad enough. Here are some of the areas they do not include: "related chemicals" such as products like sulfur and soda ash, hydrocarbon feedstocks, semifabricated plastics, and photographic formulations; petroleum; primary metals of all sorts; food; clay and glass; pulp and paper; coal and coke; and electronic components. The industries producing these products rely considerably on chemical technology, even though they are considered to be nonchemical industries. Together they employ more than half of all the chemists and chemical engineers working in industry. The larger group formed by adding all these to Chemicals and Allied Products is called the *Chemical Process Industries.*

Finally, substantial numbers of chemists and chemical engineers are employed by companies even outside the chemical process industries, such as General Motors and IBM. Their products are in no way chemicals, but a vast amount of chemical technology is required for the development and support of these products.

Thus the segment of industry we would like to talk about is very broad indeed. In most of what we say, we will try to include all these areas where chemists and chemical engineers are likely to be employed. In our illustrative examples, however, watch for the phrases "industrial chemicals," "chemicals and allied products," or "chemical process industries," which will signify only the narrower categories defined above.

HOW IT WORKS

Let's amplify the definitions above in an effort to establish exactly what the chemical industry does. What are its products, who are the big companies, and what role does it play in our economy?

Several good books on the chemical industry have appeared recently, and we find ourselves faced with the task of attempting to extract a small amount of useful information out of a vast reservoir, not the easiest thing to do. For further reading, we particularly recommend the American Chemical Society publication *Chemistry in the Economy* (1) and the Manufacturing Chemists Association's *The Economics of the Chemical Industry* (2). Our comments in this section follow the outline of the *Kline Guide to the Chemical Industry* (3). For additional facts and figures we have referred to the MCA's annual *Statistical Summary* (4) and the annual "Facts and Figures" issue of *Chemical & Engineering News* (5).

OBJECTIVES AND PRODUCTS

As briefly as possible, then, here's what the American chemical industry is all about: It starts with raw materials which were once abundant, but about the supplies of which we worry more and more these days. They are obtained from mines, forests, sea, air, farms, and wells. The chemical industry applies processes which transform these chemical substances in 12,000 plants into more than 10,000 chemicals, such as acids, alkalis, salts, organic compounds, solvents, gases, and colorants. These intermediate products are used in two basic ways.

First, the chemicals and allied products industry uses these chemicals to produce cosmetics, detergents and soaps, drugs and medicines, dyes and inks, explosives, fertilizers, paints, pesticides, plastics, fibers, rubbers, and many others. And second, other industries use the chemicals in the production of aircraft, building materials, electrical equipment, machinery, automobiles, and many other durable goods; and such non-durable goods as beverages, food products, leather, paper, petroleum products, textiles, and many more.

Ultimately, of course, these products satisfy such fundamental human needs as health, food, clothing, shelter, transportation, communication, and defense.

In general, we can distinguish three types of chemical products: basic chemicals, such as sulfuric acid and xylene, which are used for industrial processing or for the manufacture of other chemicals; goods such as synthetic fibers and elastomers, which are fabricated into end products by other industries; and chemical end products, such as fertilizers or antifreeze, used directly by the consumer.

CHARACTERISTICS OF THE BUSINESS

To effect the transformations which turn raw materials into end products, the chemical industry uses two basic technologies: formulation, the very old process of mixing things together to produce a desired result; and synthesis, the much newer technique of producing new chemical compounds, developed only after the rise of science out of alchemy. Synthesis has made the chemical industry what it is today, contributing large numbers of new products in the years since about 1930. During that period the chemical industry has grown steadily, at a rate of about 10% per year.

This growth has been sustained by high investment in research and development, providing both new products and new efficient manufacturing processes. The processes have required high capital investment: The money invested in plants and equipment per amount of product sold is higher in the chemical industry than in any other manufacturing industry save its own branch of petroleum refining.

The combination of high growth rate with new products and new processes has led to an intensely competitive industry, in which many companies account for its total production. Continuously falling prices

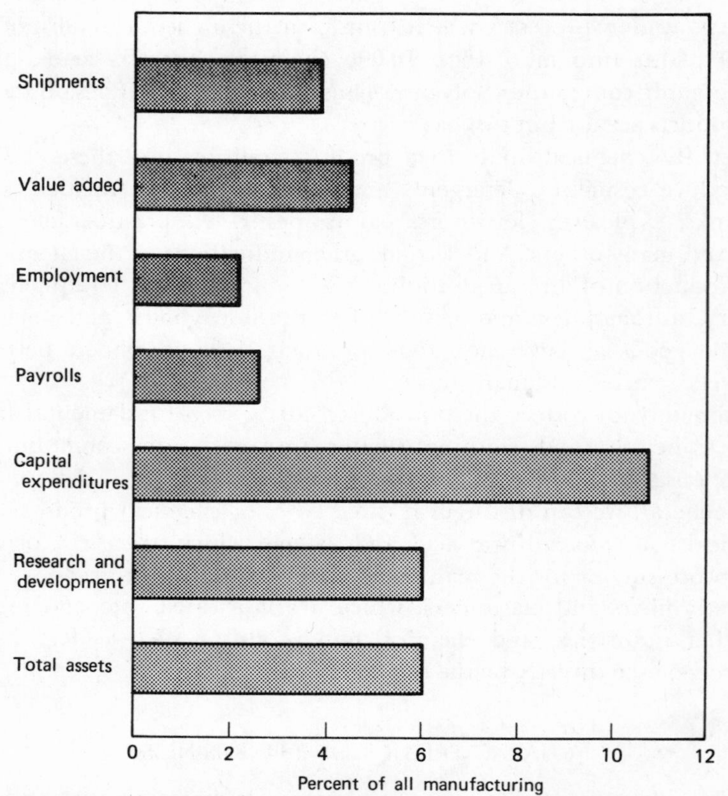

FIGURE 5. Industrial chemicals as a percentage of all manufacturing in 1967 (2). (Reprinted by permission of Jules Backman, author.)

have characterized the industry. For many years, they could be supported by gains in productivity, but in the last decade or so this approach has proved inadequate, and profit margins have declined.

As a whole, the chemical industry accounts for a very significant part of all manufacturing. As Fig. 5 shows, even the "industrial chemicals" segment of it is substantial. The figures are for 1967, but do not change much from year to year.

THE COMPANIES

In rough figures, the top four companies in the chemical industry account for 25% of total sales, and the top ten for about 35%. Although these figures may seem at variance with our earlier statement that many companies instead of a few contribute to the chemical industry, there is no contradiction. Compare these figures to those for automobiles, aircraft, tires, glass, and copper, where well over 80% (for glass, 99%) of sales is accounted for by the top eight companies. In the chemical industry, the top ten American companies don't change much from year to year. They are listed in Fig. 6, with an indication of their total and chemicals sales volume in 1972.

Figure 6 shows also that the ratio of total to chemical sales varies greatly from company to company. Before 1940, one could always tell a chemical company—it sold virtually nothing but chemicals. This is no longer true. In recent years, as diversification has become more popular,

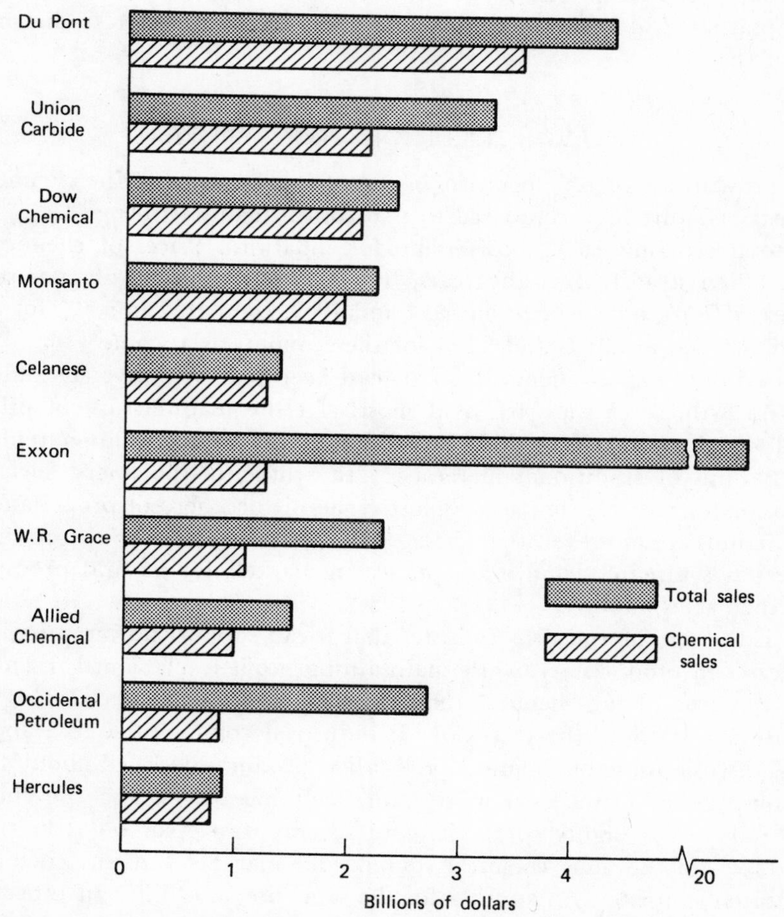

FIGURE 6. Total and chemical sales in 1972 of the top ten companies in terms of chemical sales (5).

some chemical companies have added other types of products, and other companies have moved into chemicals. Now only about a third of the top 50 to 100 "chemical" companies do more than half their business in chemicals, but they account for over half of the chemicals sold by this group. Other important types of companies producing chemicals include petroleum, natural gas, metals, food and beverages, machinery, and pharmaceuticals.

Forward integration, to provide for utilization of chemical intermediates, and backward integration, to ensure sources of raw materials, appear to provide the major reasons for this diversification, but another objective is to maintain competition between established product lines and newer materials. For example, both rubber and steel companies have diversified into plastics, which they see as a long-term threat to their traditional products.

GROWTH, PRICES, AND PROFITABILITY

The growth rate of 10% per year in the physical output of the chemical industry is quite high compared to that for all manufacturing (4.4%) or the total economy (6.9%, correcting for inflation). Prices of chemicals have fallen steadily over the years, but the dollar volume of sales still increased 7.5% a year between 1954 and 1969, compared to 5.8% for all manufactured products and 6.5% for the economy as a whole.

One would expect chemicals to at least keep pace with average industrial growth, since they are used mostly in the manufacture of other products. The extra growth has come from new products, either replacing natural or traditional materials with synthetics (polymers, surfactants), or for entirely new uses (selective herbicides, aerosol propellants). Predictions are, however, that the growth rate will slow down (in normal times) as synthetics achieve acceptance in markets for natural products and their sales stabilize.

It is the sign of a healthy industry that it can support falling prices by increases in productivity, while maintaining profit levels. Until recently, this was true of the chemical industry. Over the last 30 years, a highly inflationary period, the prices of all industrial commodities rose about 32.5%, while those of chemicals and allied products declined about 4%.

Competition furnishes a motive for declining prices, and the chemical industry is highly competitive in several ways. Not only do rival producers of the same chemical compete for markets, but one chemical can often compete with another for the same use, and different processes can compete in the production of the same chemical.

Two reasons have existed for the support of falling prices. First, most of the cost of chemical plants lies in machinery and equipment; the incremental cost of producing more product (within the plant's capacity) is small. Thus to some extent an increase in volume can accommodate a lower price; or conversely, a decline in price creates a larger market. Second, the high degree of technical competence within the industry has given it an excellent record of increases in productivity via new products, new and improved processes, and larger, more efficient plants.

All this is typical of a dynamic situation, with instances of too many producers and price decreases, but also of much research and development effort leading to new products, new processes, and new markets as well as new producers. Profits depend on the ability to reduce costs faster than prices.

Profitability in the chemical industry fluctuates with the economy, but from the late 1940s to the late 1960s, chemical companies showed consistently better performance than all manufacturing. Since 1966, the trend has been down somewhat, though there are large differences from one segment of the industry to another. With the extent and consequences of the energy crisis presently unknown, it is very difficult to predict how profitability may change in the future. Indeed, such a statement can be made about almost any aspect of life, let alone the chemical industry, these days. We can only hope that human adaptability and technical ingenuity will find means for the survival of life styles not too different from what we have known, so that the picture of the chemical industry developed here will not be anachronistic when you read these words.

SIZE

There are many measures of the size of the American chemical industry, and by any of them it's big. Some we have mentioned briefly before; here are a few more.

VOLUME OF BUSINESS

Table 1 indicates the size of the chemical processing industry in relation to all manufacturing industries and to all industries, including service, etc. Figures are given for 1950, 1960, and 1970 to indicate growth rates. Note the rapid growth of the chemical industry in the last decade, and its very substantial contribution to the manufacturing and industrial economies. Another measure of this contribution is to note that half of the top chemical companies listed in Fig. 6 are in the top 50 industrial companies of all kinds in the United States.

TABLE 1

Income and Employment in the Chemical and Other Industries*

	Income ($ billions)		
	1950	1960	1970
All industry	240	415	760
Manufacturing	76	126	218
Chemical processing industry	30	49	79
	Millions of employees		
	1950	1960	1970
All industry	49	57	72
Manufacturing	15	17	19
Chemical processing industry	5	6	13

* Excerpted and rounded off from (1).

A number of other measures of the size of the industry are given in Table 2. These figures are for chemicals and allied products, a much smaller segment of the entire chemical industry, as can be seen by comparing the income figure with that in Table 1.

Finally, one of these indicators, income, is compared for a number of parts of the greater chemical industry in Table 3.

PATENTS

Still another sort of indicator of the importance of the chemical industry is the number of U.S. patents issued for chemical compositions and

TABLE 2

Indices of Chemical and Allied Products Industry Size*

Index	Value in 1972 ($ billions)	% Increase 1972 over 1962
Sales	70.0	130
Income (after taxes)	4.5	105
Assets	61.0	124
Capital expenditures	3.5	121
Research and development funds	1.7	81

* Excerpted and rounded from (4).

TABLE 3

1973 Income of Various Parts of the Chemical Industry*

Classification	Net income in 1973 ($ billions)
All manufacturing	48.1
Chemicals and allied products	5.7
Basic chemicals	2.6
Drugs	1.5
Petroleum and related	7.4
Rubber and related	1.1
Primary metals	3.0
Stone, clay, and glass	1.3
Paper and related	1.4

* Excerpted and rounded from (5).

processes. This number was a small and relatively constant 2–5% of all patents until the early 1920s, but has increased drastically since then, exceeding 20% in the last decade as the result of vigorous gains in plastics, fibers, drugs, pesticides, and petrochemicals (1). Patents are considered more fully in Chap. 10.

MANPOWER

Although manpower utilization in the chemical industry is of major interest to the readers of this book, and is discussed more fully later, the number of chemists and chemical engineers employed in the United States is not at all a good measure of the importance of American chemical industry. The reason lies in the unusually high productivity per worker in the chemical industry. Only about 190,000 professional chemists and chemical engineers were employed in industry, government, universities and colleges, and nonprofit institutions in 1970. (The split is about 140,000 chemists and 50,000 chemical engineers.) As Fig. 1 showed, only about ⅔ of this number was employed in industry.

The figure of 190,000 represents only about 1.7% of all professional and technical personnel. This small fraction is not directly comparable with the much larger fractional contribution of the chemical industry to all manufacturing, for example, because of the great employment of professionals in nonmanufacturing fields such as medicine. Nevertheless, the conclusion seems inescapable that chemists as a whole contribute to the nation's economy in an amount far greater than that indicated by their numbers alone.

LOCATION

The geographic locations in which the American chemical industry is concentrated are clearly important in determining where the prospective employee is likely to find a job. Most figures (3, 4) deal with volume of business, but those reflecting employment are more pertinent. Both are shown, for the country divided into nine regions, in Fig. 7. The business figure most commonly quoted is value added by manufacture, roughly the difference between sales price and raw materials cost. Only total

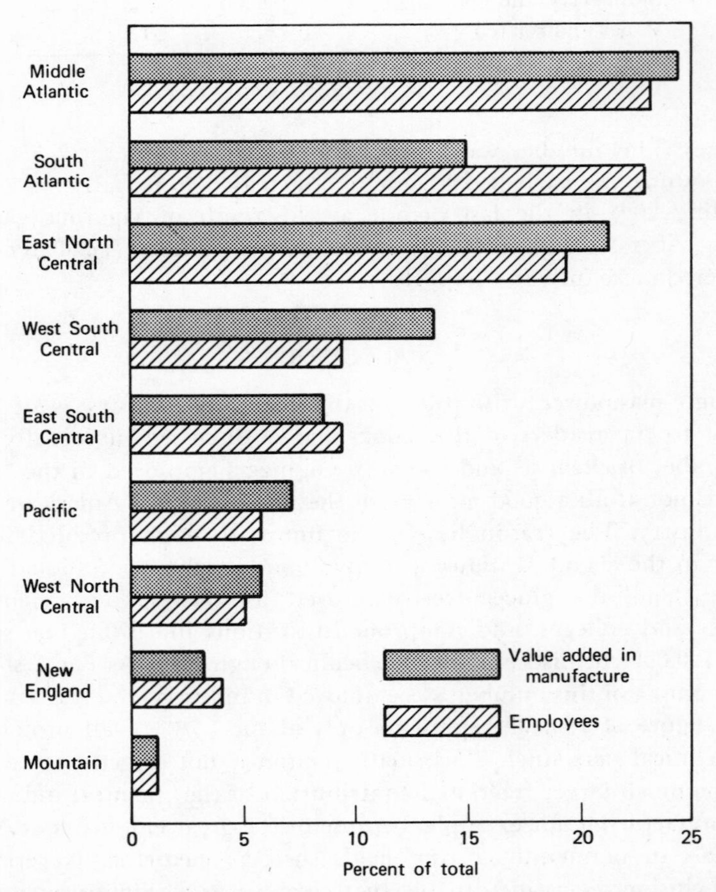

FIGURE 7. Geographic regions in which the chemical industry is concentrated, rated by value added during manufacture and by number of employees. Figures for 1971 excerpted from (4).

employment figures are available, including technical, wageroll, and clerical workers.

The Middle Atlantic states (New Jersey, New York, Pennsylvania) are still the leading region in both business and employment. Much of this business is in smaller volume and specialty chemicals such as fine chemicals, intermediates, and formulated products.

The East North Central states (Illinois, Indiana, Michigan, Ohio, Wisconsin), traditionally heavy manufacturing states, are second in business but third in employment. Second place in number of employees is held by the South Atlantic states (Delaware, D.C., Florida, Georgia, Maryland, the Carolinas, and the Virginias). This reversal probably reflects the lower wage scales for unskilled labor that have induced many companies to relocate from New England to the south. These states are the major producers of synthetic fibers, fertilizers, and gum and wood chemicals.

The position of the West South Central states (Arkansas, Louisiana, Oklahoma and Texas) results from the Texas and Louisiana oil fields, providing feedstocks for basic and intermediate organic chemicals and polymers, as well as the petroleum industry.

The data in Fig. 7 do not fully reflect the employment opportunities for technically trained chemists and chemical engineers, particularly those going into research and development rather than manufacturing. While appropriate figures are not readily available, it is our feeling that the concentration of research and development employment is much higher, across the board, in the Middle Atlantic and East North Central states than Fig. 7 indicates. There are certain obvious exceptions: If you go with a textile firm, you are quite likely to find yourself in the South Atlantic region; with a petrochemicals company, in the West South Central, and so on. But many of the large companies still do most of their R&D at the home sites (Du Pont in Delaware, Dow in Michigan, etc.) even though large manufacturing plants are located throughout the country.

MANPOWER

We shall consider employment of professionals in the chemical industry under three headings: *supply,* the total number available, and the production of new professionals in universities; *usage,* in the chemical industry as compared to others, and within the industry by field and by academic discipline; and *demand,* again by discipline, as indicated by R&D spending, and as projected for the future.

SUPPLY

Figures for the total number of chemists and chemical engineers that would be available to industry are somewhat uncertain. The American Chemical Society (1) estimates it to be about 200,000, not much above the figure of 190,000 mentioned below (and earlier) for total employment in these two categories. But the supply would no doubt fluctuate with demand: If the need were great enough, chemists now working in other areas such as medicine, women who have left the profession temporarily for family reasons, and others could easily cause this number to increase significantly.

Of greater interest to young professionals, perhaps, is the question of turnover in the supply. How many are coming out of colleges and universities with degrees in these fields? How does this compare with the attrition rate due to retirement, change of field, or other reasons? In short, will the supply exceed, just equal, or lag behind the demand?

Leaving the answers to some of these questions to the latter part of this section, we see in Table 4 that there has been a great increase in M.S. and particularly Ph.D. degrees in the last decade. Since these graduate degrees cost more, take more time, and imply far more in the way of professional commitment, the figures require a closer look.

The increase in Ph.D. output occurred primarily in established programs, rather than in new schools, although in the same time period (1960–1970) the proportion coming from the south and southwest increased markedly compared to that from the northeast. There seems to have been no significant compromise of standards involved. Heavy foreign enrollment, up to 20% in Ph.D. programs, has played a part.

Projections for the 1970s suggest a further increase of 50–125%, but

TABLE 4

Degree Production in Chemistry and Chemical Engineering, 1950–1970*

		Production in academic year ending		
Discipline	Degree	1950	1960	1970
Chemistry	B.S.	10,600	7,600	11,600
	M.S.	1,600	1,200	2,100
	Ph.D.	950	1,050	2,150
Chemical engineering	B.S.	4,500	2,950	3,700
	M.S.	700	600	1,050
	Ph.D.	180	170	440

* Excerpted and rounded from (1).

this is probably highly optimistic. Support has fallen off drastically (save for strong federal support of health-related programs in biochemistry), and it is far more likely that the number of Ph.D.'s will decline rather than increase over the next few years. The trend has already started, and it may be that production of Ph.D. chemists will stabilize at about 1500 per year for the rest of the decade. This is probably quite consistent with the demands of the job market. In chemical engineering, however, there is already a significant shortage of new Ph.D. holders.

USAGE

We referred earlier to the figure of about 190,000 professional chemists and chemical engineers employed in the United States in all respects. The ACS (1) states that about 70% of these are employed in private, for-profit industry. This percentage has remained rather constant for the past 20 years. Of the remainder, some 17% are employed by universities and colleges, 9% by the government, and the remainder in nonprofit institutions in 1970.

Turning now to the 140,000 or so in industry, Table 5 shows how they are approximately divided among the several types of work in that sector. The figure of 47% in research and development is consistent with other surveys. It varies somewhat with the size of the company, however, reaching 50–55% in small companies, somewhat less in very large ones.

Let us now look at how the employment is divided among the various segments of the American chemical industry. Table 6 shows 1973 figures, but those for 1972, consistent with our other data, are not significantly

TABLE 5

Primary Activities of Industrial Chemists, 1970*

Primary activity	Percent engaged	
Research and development	47	
Basic research		19
Applied research		17
Development		11
Management or administration	32	
R&D		17
Other		15
Production and inspection	16	
Other	5	

* Computed from data in (1).

TABLE 6

Employment by Segment of the American Chemical Industry, 1973*

Industry segment	Percent employed	
Chemicals and allied products	40	
Industrial chemicals		12
Plastics materials and synthetics		9
Drugs		6
Soap, cleaners, and toilet goods		5
Paints and allied products		3
Agricultural chemicals		2
Other chemical products		3
Petroleum and coal products	7	
Rubber and plastics products	26	
Paper and allied products	27	

* Calculated from data in (5).

different. The data are, however, for all employees instead of professionals only.

Finally, it is without doubt interesting to young professionals to see how the employment of chemists is related to fields somewhat approaching the traditional academic disciplines. As Fig. 8 shows, the four classical disciplines account for substantially less than half the total, and the addition of biochemistry scarcely changes the picture. Polymer chemistry, agriculture and food chemistry, and a number of other areas not emphasized in the university make up the remainder. Small wonder that industrial employers are anxious to hire well-rounded, adaptable, flexible candidates, as emphasized in Chap. 3.

DEMAND

Because of the unavoidable time lag between our writing this book and your reading it, it is difficult to make relevant statements about the demand for professional chemists and chemical engineers. We can say that toward the end of 1974, the job outlook was still good (7). Employment was on the rise, unemployment of professional and technical personnel was at about 2%, its lowest level in four years, and the volume of job advertising had been rising for two years. Industrial spending for research and development, another measure of demand, was expected to increase beyond the erosion of inflation in 1974. There seemed to be plenty of jobs for chemists, and a real shortage of chemical engineers

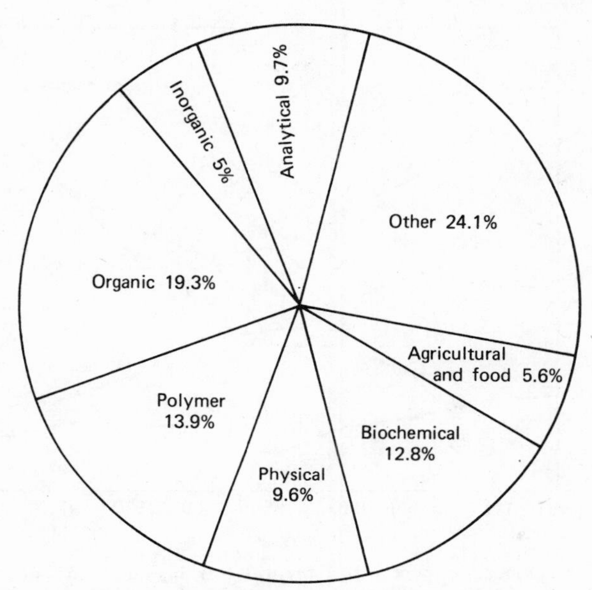

FIGURE 8. Distribution of industrial employment of chemists by field (6).

appeared to be developing.

Recent past performance is also of interest in relation to employment opportunities as a function of academic discipline. Figure 9 shows clearly the trend toward fewer opportunities in inorganic and physical chemistry, the approximate status quo in organic, and the rises in analytical chemistry (slow) and biochemistry (spectacular).

But these words are of necessity out of date when written. To be more timely, one must resort to forecasts, all of which are based on some scenario of national and world development which seemed plausible when the forecast was made. Unfortunately, none of the scenarios used to date include the energy crisis and its worldwide resounding implications. So we shall quote them only briefly, and with tongue in cheek.

The ACS (1) quotes a Bureau of Labor statistics projection, and the results of a survey among its own corporate associates. The latter, more conservative, seems to us to be at the same time more realistic. It calls for an average growth rate of professional employment in the chemical industry of just under 2% per year through 1980. Translated into jobs, this means about 2700 new openings per year. To this must be added about 2200 for replacement; in round numbers, about 5000 new jobs per year for chemists and chemical engineers, B.S., M.S., and Ph.D.

Clearly, if B.S. production were to continue at the same rate as indi-

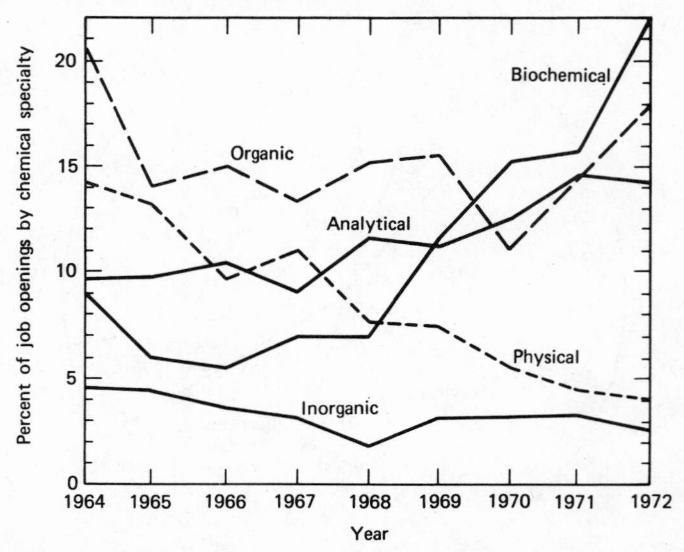

FIGURE 9. Employment opportunities for the various academic disciplines, 1964–1972 (8). (Reprinted by permission of *Chemical & Engineering News.*)

cated in Table 4, and 70% of these people had to be absorbed into industry, we would be in trouble. This is not likely to be the case, both because of decreasing college enrollments and because far from all these graduates will go into the chemical industry: many go into biochemistry, the life sciences, material sciences, medicine, or the general (nonscientific but equivalent to professional) labor force.

We close by repeating the warning that predictions aren't likely to be of great value in drastically changing times. It could be that the needs of our battered environment plus the requirements for the development of new sources of energy will drastically increase the demand for chemists. Or, an extended recession could have the opposite effect. Our crystal ball is too cloudy to show the answers.

SALARIES

In the area of pay, we will shun predictions, being unable to cope with inflation today, much less to outguess it for the future. The stage for this section was set in Chap. 1, where we cited the relative positions of chemists and engineers among other classes of white-collar workers, and in Fig. 4, where we made similar comparisons with other branches of science. Here we shall consider only how salaries vary, and what the trends are, within the chemist-chemical engineer family.

The trend in median starting salaries for chemists in industry is shown, by degree level, for the period 1968–1974 in Fig. 10. Despite some fluctuation at the B.S. and M.S. levels, the overall trend is up. What is more important for the longer term, the salaries for experienced chemists are also up at all degree levels, by amounts which substantially exceeded inflation, at least up to 1974. Over the decade 1962–1972, for example, the average annual gain for all chemists was 6.5%—slightly higher for B.S. and lower for Ph.D.'s. Gains in 1972–1974 exceeded this rate significantly. It is interesting to note that, although the salary gap between holders of B.S. and M.S. degrees eventually narrows and virtually disappears as years of experience mount up, the difference (in dollars) between Ph.D. holders and the others does not (Fig. 11).

Median starting salaries for chemical engineers are usually somewhat higher than for chemists, at all degree levels. Data for the B.S. level are shown in Fig. 12 as a function of geographic location, which demonstrates some minor regional differences. The figures for chemists are somewhat lower than those in Fig. 10 because they include academic as well as industrial salaries.

The ACS survey on which these data are based also showed that the median industrial starting salaries for women are significantly higher

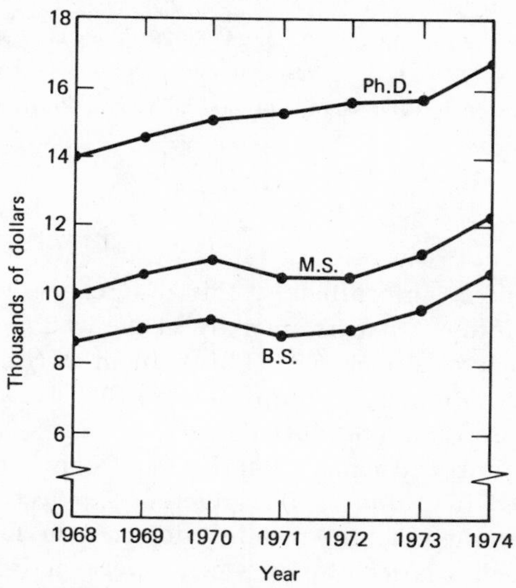

FIGURE 10. Median starting salaries of chemists at various degree levels, 1968–1974 (9). (Reprinted by permission of *Chemical & Engineering News.*)

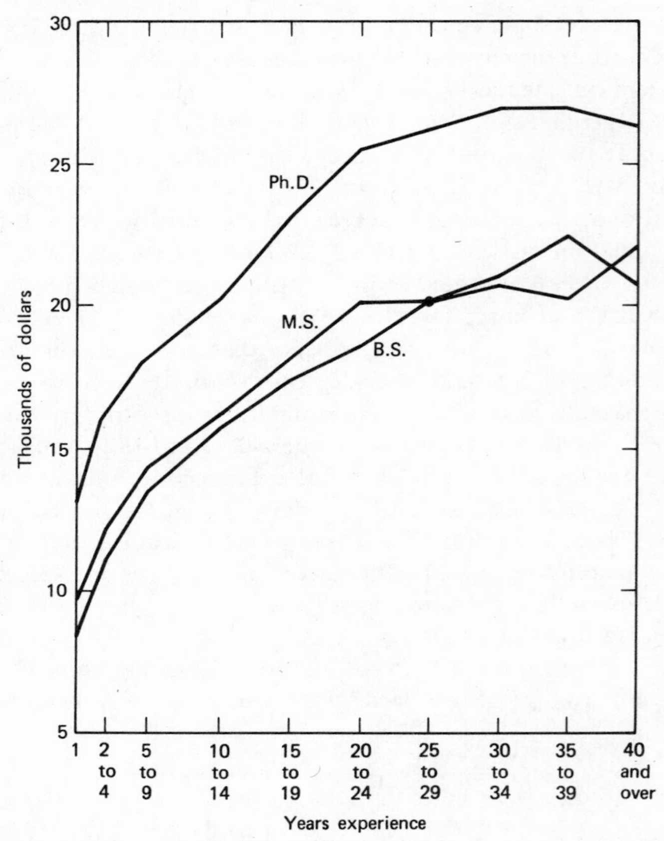

FIGURE 11. Increase in median salaries of B.S., M.S., and Ph.D. chemists with years of experience (10).

than those for men. Unfortunately, this difference—which is novel—doesn't hold up: After 2–4 years, men are in the lead, and overall, make half again as much as women (10). This gain in starting salaries is not yet seen in respect to minority groups, though they are as a whole better qualified (larger proportion of Ph.D. degrees).

Within the industry, chemists' salaries vary significantly by field of work, as indicated in Table 7. The differences reflect the demand for chemists with certain kinds of training depicted in Figs. 8 and 9. As might be expected, salaries also vary with work activity, as shown in Fig. 13. Teacher's salaries have been included as an additional reminder that industry pays better than academia.

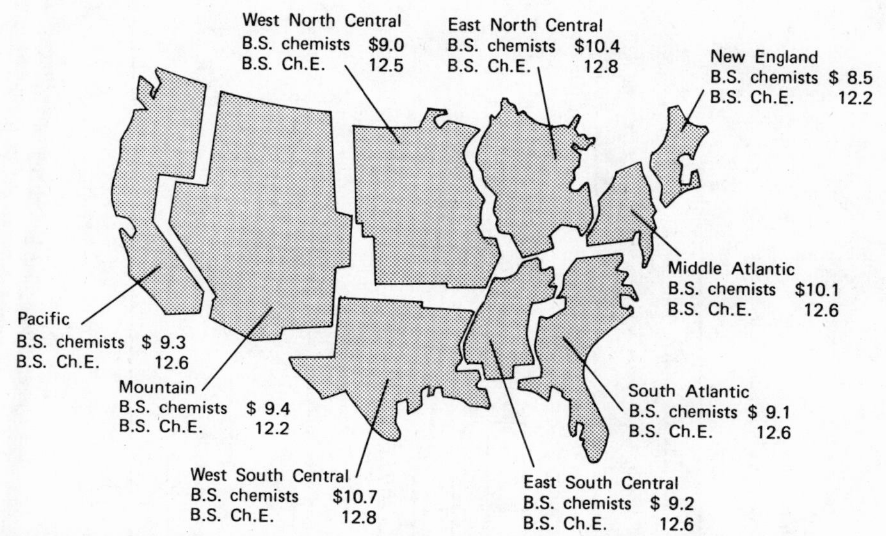

FIGURE 12. Median starting salaries of chemists and chemical engineers as a function of geographic location, 1974 (9). (Reprinted by permission of *Chemical & Engineering News.*)

TABLE 7

Chemists' Salaries by Field, 1973*

Field	Median annual salaries (thousands)		
	B.S.	M.S.	Ph.D.
Analytical chemistry	$15.6	$17.0	$19.5
Biochemistry	16.0	15.5	21.0
Inorganic chemistry	16.2	16.0	17.0
Organic chemistry	17.0	18.0	20.3
Physical chemistry	17.0	18.0	20.4
Polymer chemistry	18.0	18.6	22.4
Other specialties	18.0	18.6	22.0
All chemists	16.8	17.5	20.5

* Adapted from (10). Reprinted by permission of *Chemical & Engineering News.*

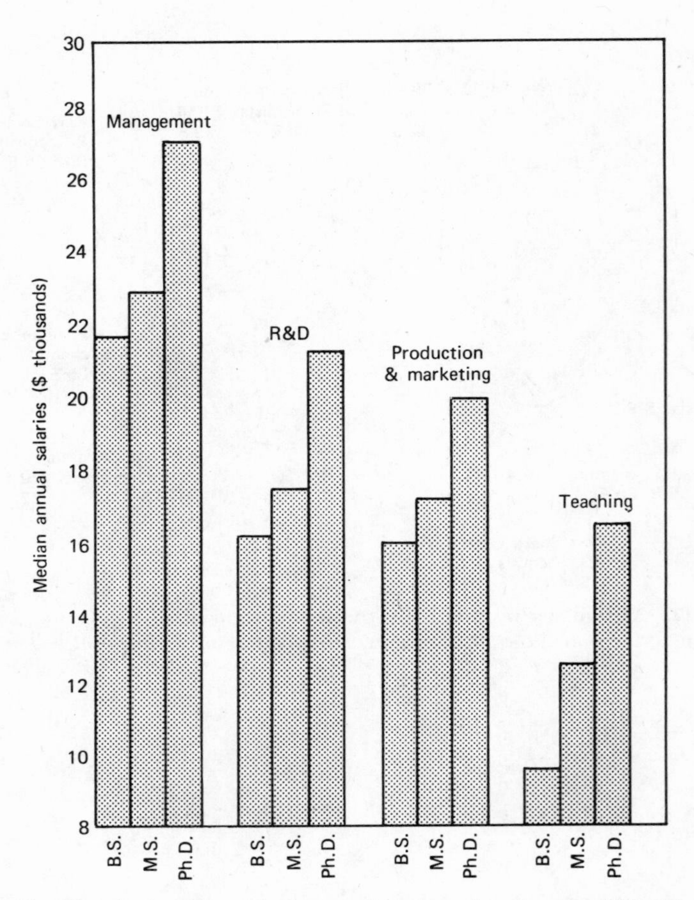

FIGURE 13. Chemists' median salaries as a function of work area, 1973 (10).

TABLE 8

Chemical Engineers' Salaries According to Number of Employees in Company, 1973*

Number of employees	Median annual salary ($ thousands)	Percent of companies
1–199	18.5	15.3
200–999	18.0	21.5
1000–4999	19.5	24.2
5000 and over	20.0	39.0

* Adapted from (11). Reprinted by permission of the American Institute of Chemical Engineers.

Finally, a survey by the American Institute of Chemical Engineers (11) shows that chemical engineers' salaries are greater in larger companies, as indicated in Table 8.

ORGANIZATION

While we defer detailed discussion of the organization of chemical companies until Chaps. 6–9, a few words are not out of place here. We've noted before that chemical companies come in all sizes, from very small to exceedingly large. But once a certain minimum size has been reached, getting them out of the "one man, garage" operation, the basic organization of the company doesn't change much. Rather, it first proliferates as backup levels of management and support groups are added, then often splits into parallel operations involving different (though sometimes related) products or processes, in order to keep the structure from being too unwieldy.

In a very small company, virtually full attention must be paid to making the product and getting it out the door. Thus production is the first function to be staffed. Soon in the growth pattern, however, attention must be paid to such auxiliary responsibilities as purchasing, inspection and quality control, shipping, and the like.

The next major functions to be considered for separate staffing are sales and product improvement. Often a small company will utilize the services of a firm of sales representatives to take its product to the marketplace, rather than hire its own salesmen. This is common to other fields too, such as small scientific instruments. As sales grow and customer needs are better defined, a marketing organization can be added, replacing or supplemented by the "reps."

Product improvement starts with the production staff, but as company size increases, a separate function becomes necessary. This often starts as a plant assistance group, progresses to "development," "research and development," and "research." Only when the research establishment is divorced from the manufacturing plant in location as well as organization is it likely that the last traces of on-the-spot plant assistance on demand will disappear.

An interesting article on small chemical companies (12) points out that the organization, based on this general scheme, remains simple and streamlined until the company's assets (value of plant, materials going through the pipeline, etc.) reach about $10 million. Beyond this point the company must begin to build management in depth, developing second lines to back up its key executives and adding staff groups to perform various functions (legal, patent, information retrieval and library, safety, and market research, to name only a few). At this point the company begins to lose its flexibility and aggressiveness, as decision making becomes slower and more cautious. By the time company assets have reached about $25 million, it functions as a small corporation, and a glance at the headings for Chaps. 6–9 will give clues to the intricacies of its organization.

It is about at this point in size, too, that the company begins to require the services of recent graduates, whom it can train to perform the various tasks required of its research, development, production, and support groups. The very small company can seldom afford the luxury of hiring a person largely unskilled in industry and supplying the necessary training. These companies will look for personnel from other, perhaps larger, firms who are challenged by mode of operation and growth potential of the small outfit. We expect that most of the readers of this book will have primary interest in the larger rather than the very small companies.

At the other end of the scale of sizes, the very largest companies find it necessary to subdivide their activities, usually along product or process lines, into divisions of more reasonable size and unity of interest. The Du Pont Company, with which we are familiar, has several operating departments whose names are generally indicative of their major product lines: Plastics, Textile Fibers, Elastomers, Photo Products, etc. These departments are essentially independent corporations except that their chief executive is responsible to and draws direction from the overall management of the company.

Each department has its own research, manufacturing, and sales divisions, its own research laboratories, plant sites and sales offices. Occasionally facilities are shared; for example, the department may have a research laboratory at a central companywide research facility called the "experimental station." At this location there are also laboratories of nonoperating departments (those that are essentially service organizations, making no products), concerned with engineering or chemical research. Other nonoperating departments (legal, secretary's, etc.) also provide services for the operating departments as well as the company as a whole.

This type of organization is advantageous to the company because it preserves some of the flexibility of decision-making characteristic of smaller companies, though the free-wheeling and often more daring policies of the very smallest are not achieved. It is advantageous to the employee because he can take a well-defined part in an organization whose size is not overwhelming to him.

MULTINATIONAL COMPANIES

Although this book is primarily concerned with the American chemical industry, we can't overlook the importance of a new phenomenon in the last 10–15 years, the so-called multinational companies (13). Hard to define, these concerns are nevertheless the key element in a force that is tying industrialized parts of the world together into one economy that tends to transcend the nation-state concept in a way that politics has been totally unable to do. The role of chemistry in world economics is no less important than it is in U.S. economics, and the participation of the United States in the world market is highly significant.

The concept of multinational companies is so new that there is no consensus on the definition of the term or what is encompassed by it. One set of criteria is that such a company does business in many countries of varying degrees of economic development, has foreign subsidiaries with the same sort of organization we have discussed before, run by nationals of that country, has a multinational headquarters staffed by people from the different countries involved, and has stockholders in different countries.

It is hard to list such chemical companies, but there are probably about 40 of them, doing perhaps a quarter of all the chemical business in the world. About a dozen of these are U.S. based, including Du Pont, Union Carbide, Monsanto, Dow, W. R. Grace, Rohm and Haas, and Hercules. The fraction of the sales of these companies originating in foreign operations ranges from about 20% to almost 50%.

The growth of a company to multinational scope appears to start, typically, with an export business and possibly some licensing of technology to overseas companies. As foreign sales grow, some fabricating or finishing operations may be set up in the countries where business is best. Most companies sooner or later establish an international division to handle their overseas business if it grows large enough.

But true multinational operation demands much more. A change in perspective from national to global is required, and must be accompanied by adjustment of the corporate structure to a global stance. The appendage international division is replaced by a structure that places operation in all countries on a potentially equal footing. Organization may be in terms of geographic locations, so that the management group responsible for North American operations is on a par with its European or Asian counterparts. Alternatively, the organization may be along product lines, so that management for each product area is responsible for it everywhere in the world.

The implications for the technical employee, particularly as he advances into management, are obvious. As a member of a multinational company, he has an unparalleled opportunity to live and work overseas, shoulder to shoulder with his colleagues from other countries. To some, this will be an exciting and challenging prospect.

THE FUTURE

Predictions are difficult enough in normal times, and as we write, times are far from normal. We can only hope that our accustomed mode of life will survive without too drastic a change the onslaught of energy and ecology crises, and will base our remarks here on the assumption that this will be the case. A recent article (14), based on the same assumption, predicts that by the end of this century the "chemical industry" will no longer be recognized as such. Pointing out the difficulty of defining the industry as we did at the start of this chapter, the author postulates that one will then speak in terms of "synthetic materials," "fossil fuels and hydrocarbon materials," and "pharmaceuticals and cosmetics" industries, with whatever else is now considered part of the chemical industry distributed appropriately elsewhere. We think this makes good sense.

By whatever name we call it, the more important aspect is what will happen to these new, or the present, industries. The author predicts continuing growth, though at a somewhat smaller rate than in the past, say 7% per year instead of 10. We are not expert enough to judge the correctness of his figures, but we agree that continuing growth is likely, regardless of the effects of energy and ecology crises.

If we concede that many of man's present problems are caused by technology, we nevertheless insist that the abandonment of technology won't solve them. What is needed is a redirection of efforts toward the solution of these problems. This is also predicted in the article cited, as a reorientation of the industrial research effort toward the problems of building, transportation, communication, and other environmental and urban problems.

If such a change comes, there will still be urgent chemical problems requiring solution, including resource conservation and recovery, energy generation, and nutrition, but manpower needs will still change. Industry will require scientists and engineers with a broader spectrum of training and greater versatility, to adapt to changing goals.

Here we note that this requirement is really nothing new, for breadth and versatility are just what industry has been asking for all the time.

So we shall put the cover back on our cloudy crystal ball and address, in the next chapter, the questions of what industry wants from you, and how you can land the job which will show the company what you can do.

REFERENCES

1. *Chemistry in the Economy*, American Chemical Society, Washington, D.C., 1973.
2. Jules Backman, *The Economics of the Chemical Industry*, Manufacturing Chemists Association, Washington, D.C., 1970.
3. Patricia Noble, ed., *Kline Guide to the Chemical Industry*, 2nd ed., Kline, Fairfield, N.J., 1974.
4. *Statistical Summary 1973*, a supplement to the *Chemical Statistics Handbook*, 7th ed., Manufacturing Chemists Association, Washington, D.C., 1973.
5. Anon., "Facts and Figures—The U.S. Chemical Industry," *Chem. & Eng. News* **52** (23), 23–52 (June 3, 1974).
6. David A. Roethel and Charles R. Counts, "Realignments in Chemical Profession Continue," *Chem. & Eng. News* **49** (47), 90 (Nov. 15, 1971).
7. Anon., "Demand for Chemical Professionals Is Still on the Rise," *Chem. & Eng. News* **52** (40), 23–25 (Oct. 7, 1974).
8. Anon., "Employment Outlook 1973," *Chem. & Eng. News* **50** (40), 9 (Oct. 2, 1972).
9. Anon., "Salaries Are on the Rise at All Experience Levels," *Chem. & Eng. News* **52** (40), 28, 32 (Oct. 7, 1974).
10. Anon., "Employment Outlook Brightens for Chemists," *Chem. & Eng. News* **51** (25), 6 (June 18, 1973).
11. F. J. Van Antwerpen, letter to members, American Institute of Chemical Engineers, 1973.
12. Walter S. Fedor, "Small Chemical Companies—A Constant Challenge to Grow," *Chem. & Eng. News* **44** (24), 136 (June 13, 1966).
13. Anon., "The World Chemical Economy," *Chem. & Eng. News* **51** (16), 19 (April 16, 1973).
14. David M. Kiefer, "Chemicals 1992: 20 Years of Industrial Change," *Chem. & Eng. News* **50** (28), 6 (July 10, 1972).

CHAPTER 3

LANDING THE JOB

We have described the American chemical industry and its manpower needs; we now come to you, the young professional, and how you go about making your bid in the job market. Consider these questions: do you have the personal qualifications that industry is looking for? How do you contact the employers with job openings? . . . sell yourself to the potential employer through your personal résumé and at the campus interview? . . . conduct yourself on a plant visit to give the best impression? . . . and, finally, evaluate and select among the various companies which have (hopefully) come forth with a job offer?

In this chapter we explore these subjects and give you our advice for every step of the way, from the initial search for prospective employers to the letter of acceptance of a job offer. Some comments on changing jobs are included, too, for those whose first berth wasn't as comfortable as hoped. All this is preliminary to a discussion of your responsibilities as an employee (new or seasoned), covered in Chap. 4.

PERSONAL QUALIFICATIONS SOUGHT BY INDUSTRY

In considering this topic, we distinguish between two related areas:

- What do companies look for in selecting new employees?
- What should be the attitude of the employee in beginning a new job?

Here we concentrate on the former aspect, though both are extremely important. We pick up the latter in the section, "Advice to New Employees."

Most employers concerned with the care and nurturing of new employees express considerable concern over the adjustment and orientation that you as a new graduate must undergo when you start out with the company. They realize that generally you have had no experience in industry and often have had little enlightenment at school as to what to expect. This, of course, is well known to us and was, in part, our incentive for writing this book.

J. O. Hendricks of the 3M Company comments in a private communication that one way to prepare for the transition is to participate in a cooperative program or a summer program, in which you have an opportunity to learn about the business of the company and to become acquainted outside of your immediate study area. Obviously, anything your professor can do to gain similar knowledge and pass it on to you is highly valuable.

In addition to their common concern that you should learn as much about the workings of industry as possible (preferably by personal experience, the greatest teacher), the industrial spokesmen agree rather well on the personal characteristics they would like to see in the new employee. Here is a list developed from Hendricks' comments, a report of a survey by the American Chemical Society (ACS) (1), a short article by B. W. Rossiter (2), and a longer report by Odiorne and Hann (3).

ABILITY TO THINK

If a single attribute can be considered the most important, and if priority of attention is a measure of value, then your ability to think takes top position. This does not at all mean the ability to absorb technical knowledge; that comes much farther down the list. What is meant is the capacity for intelligent, independent thinking, exhibited by the use of logic and the display of creativity. Native intelligence is important, but it's not enough. The ability to apply that intelligence to solving a wide variety of problems, in a logical, disciplined way, is of far more concern.

Although opinions differ, there is no doubt about the importance of intelligence as measured by academic excellence. This is one indicator of your ability to think, and it's one of the most visible measures to those involved with hiring. Many companies place great emphasis on academic excellence, provided that the other characteristics in this list are also evident. They insist that it's not true that industry is looking for a well-rounded person with a C+ average. Some level of intelligence above the average is required, though it need not be that of a genius.

For these reasons, industry is most unhappy about the recent trend in some universities to allow unlimited "pass–fail" options or to do away with grades entirely. As distasteful as the competition for them is, grades are hard evidence of academic excellence, and industry wants and expects to see this evidence.

More difficult to define, and certainly far more difficult to teach, are the other aspects of independent thinking. The logical application of intelligence is an important one. Here logic means awareness and organization of facts, rather than the ability to reason. This is related to intelligence, but is a selective aspect: Which of all the facts I know is the key to solving this problem? (And in industry, the problem may have to be solved under stress.)

Creativeness is another part of the ability to think. Hard to define and even harder to measure, it is defined by some as the ability to bring out, analyze, and interpret new ideas. It's an essential part of research, but when you get up to management, it's probably more important to recognize creativity than to be creative.

ABILITY TO ACHIEVE OBJECTIVES AND GET RESULTS

If technical knowledge stands for anything in a company's judgment of a prospective employee, it is as one contributing factor to your ability to achieve objectives. But other things are important, too, including your ability to apply knowledge and a wide range of personality traits. What industry wants is the person who can discern and establish important goals and then systematically and energetically move toward their attainment. To judge this ability, the company looks for accomplishment rather than potential, for concrete results rather than just activities.

It is sometimes difficult to reconcile this ability to set and achieve objectives, which is referred to as innovation, with the deep and frequently narrow consideration of a very specific field of knowledge often required in Ph.D. thesis research. Here the versatility and sense of purpose characteristic of the innovator can too easily be lost. The value of the Ph.D. training lies in learning the discipline and methodology of science, the danger lies in getting lost in an open-ended problem or a maze of meaningless detail. Industrial companies responding to the ACS survey (1) were particularly concerned over the consequences of overspecialization in the Ph.D. program. They saw a need for modification of the program to instill greater initiative, effort on the job, and attention to continuing self-improvement in the students.

In considering this qualification, it is necessary to distinguish between technical knowledge and technical ability. The latter is the capacity to apply technical knowledge to new problems and situations. At the research level, ability usually implies knowledge, but as you advance to management levels the two separate, and ability may mean the capacity to apply broad-brush or block-diagram knowledge, leaving the details to others.

Sharing importance in this and the next qualification on the list are the two personality traits of self-confidence and optimism. Self-confidence is largely achieved through and radiated in the communication process, but as a personal rather than a communication trait, it is essential to the steady progress toward an objective. No less important is optimism; pessimism can't be tolerated in a successful business. The two go hand in hand, since the optimist is usually self-confident. It goes without saying, of course, that overoptimism and false confidence can be dangerous; part of self-confidence is honestly admitting when you have made a mistake.

ABILITY TO COMMUNICATE

There is no question in most people's minds about the importance of communication, including speaking, writing, reading, and listening. The need for skill in all these areas cannot be emphasized enough. Ideas have to be sold; they are worthless unless you can get them across to people. The successful scientist is one who has developed the ability to communicate well, both verbally and in writing. Similarly, instructions are important, and that half of communication devoted to receiving rather than imparting information should not be neglected.

The ACS survey respondents (1) were particularly critical of what they called an "amazing lack of competence in communication skills." The educators responding saw this in examination papers and thesis drafts, the industrialists in reports, articles, and proposals. All felt that better training in basic English grammar was needed far more than training in a foreign language.

While some feel that the ability to communicate, particularly orally, is an inborn quality which it is not practical to develop, we do not agree. We feel that skill comes with experience, and we encourage you to take every opportunity to practice your ability to communicate, particularly in formal presentations.

Persuasiveness, and to some extent aggressiveness, are also important here. Without the ability to persuade people to see your side of an argument, you may never see your technical knowledge applied.

ABILITY TO ADAPT

Some of the great challenges and enjoyments of an industrial career are its changing problems and objectives. The employee who can adapt to changing times and technologies—who can take everything in stride—can not only make valuable contributions, but do so with a great deal of personal satisfaction and stimulation. Adaptability is closely related to creativity, requiring imagination and the ability to create with new ideas and concepts.

It doesn't take much study of modern technological advancement to see that an interdisciplinary approach underlies much of it. Not only have the traditional disciplines within chemistry almost disappeared, but the physical sciences themselves, such as chemistry, physics, and biology, are becoming more closely interrelated. The belief in both academic and industrial circles is becoming prevalent that "those who ignore an interdisciplinary approach to science are doomed to inefficiency or irrelevancy or both . . . high performers consistently take an interest in fields contiguous to or even separate from their own discipline. Low performers seldom venture outside their own specialty" (2).

The definition of interdisciplinarity can be taken either broadly, as implied above, or with respect to areas of specialization within your major discipline. Industry is particularly critical of narrow specialization, as reported by the ACS (1), who interpret the urgent message as "send us *chemists,* not synthetic organic chemists, spectroscopists, theoretical physical chemists, etc., but *chemists*!" A similar complaint is that chemistry graduates lack "many basic laboratory skills that [they] really should have," ranging from how to read a burette to how to design an experiment.

ENTHUSIASM

B. W. Rossiter (2) says that another characteristic his company looks for is enthusiasm for the work and the energy to carry it vigorously to completion. This means a willingness, even an eagerness, to work hard. Putting in the time is not enough; it is necessary to live and breathe the job while on the job, to work at capacity, and to know when to stop lest efficiency decrease. Anyone who does not show initiative and a willingness to work hard will detract from a business—and from his chance of being hired. Aggressiveness (but controlled), persuasiveness, self-confidence, and communication skills all contribute to this essential enthusiasm.

A FEEL FOR THE INDUSTRIAL GAME

One of the least tangible, but not least important, qualifications sought by industry lies partly in experience and partly in attitude, leading to what we call "a feel for the industrial game." It helps to know what industry is all about; that's why we are testing our communication skills by writing this book. One important result of this knowledge should be abandoning the false idea (and we can see some of our more cloistered academic colleagues quivering) that industrial research is less "pure" than academic, or even second rate. Nothing could be farther from the truth: Some of each kind is good, some bad. Industry's representatives are particularly sensitive to an attitude of suspicion, superiority, or aloofness in the prospective employee—and to show a lack of interest and enthusiasm is a good way to get turned down.

What are some of the areas of industrial research that you, as a prospective employee, should try to gain familiarity with? The role of polymers is one area: Over half of the chemists and chemical engineers employed in the United States today find their work related in some way to polymers. [Modesty prevents us from doing more than mentioning one book on the subject (4).] The role of materials is another: metals, ceramics, glasses, semiconductors, and polymers (5). The industrial applications of organic chemistry to plant processes (6). The role of computers in the industrial research laboratory and the plant (7). And finally, a thorough understanding of the basic concepts of economics (8). This last should involve keeping up with the current news (9, 10) as well as book knowledge. As the ACS (1) puts it, graduates lack sufficient awareness of or respect for economic constaints, commercial applicability, and social needs or implications of their work.

PERSONAL CHARACTERISTICS

Although the list of qualifications often stops here, we can't resist adding a word on the judgment that you as a job candidate can expect to have made of your personality, and its bearing on your chances of getting a job offer. We know, from having been on both sides of the table, the importance of making a good impression at each stage of contact between job hunter and potential employer.

Personal integrity clearly tops the list of attributes that are important, despite what may seem to be the tenor of our times. Honesty, in matters of fact, and ethics, in matters of judgment, are essential. Your appearance, particularly grooming and dress, is indicative of your personal organization in a very telling way. Anything out of the ordinary can detract from your primary purpose. Social etiquette, knowledge of the proper conduct in any type of company, is also looked for with critical care. Your spouse can play an important part here. Finally, lack of some knowledge of human relations (for example, an insight into people's motivations) and of the humanities (fine arts and social science) can be a distinct detriment.

Table 9 summarizes the attributes of strong and weak job candidates as seen from the campus interviewer's point of view.

TABLE 9

What the Candidate Looks Like to the Interviewer*

The strong candidate	The weak candidate
Is intelligent, has high grades	Has been a poor student
Handles the interview well	Is awkward and ill-at-ease, unprepared for the interview
Knows what he wants in a career and why	Doesn't know what he wants to do, is mixed up but hopes for the best
Is mature and sensible	Looks and acts inexperienced and childish
Is friendly, likeable and pleasant	Has a grating personality—too brash and cocky, or hostile, cynical, or rebellious
Has been around an average amount for his age, took part in extracurricular activities	Apparently neither worked nor played much in college
Dresses normally and neatly, no major defects in appearance though not necessarily handsome	Makes the recruiter wonder whether he would like to seen in his company

* Adapted from (3). Reprinted by permission of the University of Michigan.

WHERE TO LOOK FOR EMPLOYMENT

Once the decision to seek employment in industry has been made, your first task as a prospective employee is to locate employers of interest and make contact with them. We list here the more widely used ways of doing this, in the approximate order of importance and efficiency, as we see it, for the young professional looking for his first job. Comments along the way point out the routes more often followed by the industrial employee looking for a change in jobs.

PERSONAL RECOMMENDATION

In the relatively few cases where it exists, nothing (in our opinion) equals personal assistance and recommendations by academic and industrial people who know you. If you are fortunate enough to work for or know a professor who has close contacts with industry (and how we wish there were more!), ask him for suggestions and to recommend you to the companies. If your university is at all in touch with the industrial world, you will find its scientists and engineers visiting your department from time to time. Get to know them, describe your research experience and interests to them, and write to them when you are ready to look for a job. A personal contact with or from a respected person in your field is worth its weight in gold.

PROFESSIONAL SOCIETIES

In addition to, or in lieu of, personal contacts made on campus or through your professor, membership and participation in your professional society or societies offers an excellent way of making and fostering personal contacts. Here we confine our remarks to two societies: the American Chemical Society (11) and the American Institute of Chemical Engineers (12). Student membership in each is inexpensive and well worth while. What is more important is participation in meetings, both local and national. In particular, we strongly favor the policy of having graduate students attend and present papers at regional and national meetings, sharpening their communication skills and bringing themselves to the attention of the industrial scientists in attendance.

Depending on the society, more direct benefits may accrue from attendance at a national meeting. While the American Institute of Chemical Engineers does not allow recruiting activities at their national meetings, the American Chemical Society does, and offers a variety of related services. These are described in a booklet (13) to which we have referred frequently in writing this chapter. Foremost among these is a National Employment Clearing House where direct contact between employers and candidates can be made. Members and student affiliates of the society can register for interviews at no charge. These clearinghouses are sometimes overcrowded, however, and the similar campus activity described below is likely to be more effective for the student. Local ACS sections also offer a variety of employment aids.

COLLEGE PLACEMENT OFFICE

The logical place for the student (or, in most cases, alumnus) to start the search for a job is the college placement office. Here he will find skilled personnel, a variety of literature from many companies, both large and small, and several of the directories mentioned below. A major function of the placement office is to arrange on-campus interviews between the student and campus recruiters sent out by industrial firms. This interview is of sufficient importance to be treated in a separate section to follow.

EMPLOYMENT DIRECTORIES AND MAGAZINES

There is a wide variety of written material designed to help you as a prospective employee locate the company of your choice. At the placement office you will find the *College Placement Annual* (14), a not-for-sale publication distributed free through such offices and armed forces installations. It describes employment opportunities in several thousand companies in all fields, including the chemical industry. Several other commercial directories are worth consulting: Some are aimed at the chemical industry (15); others, at industrial research (16); and still others are quite general (17). A helpful guide to job-hunting tools appeared a short time ago (18).

Several magazines and subscription services are also available. *Business World* (19) and *New Engineer* (20) are sent free to full-time or graduating students. Both these and the *College Placement Annual* contain helpful articles as well as lists and advertisements of employers. The professional-society news magazines are also helpful; in particular, *Chemical & Engineering News* (9) publishes an annual issue devoted entirely to employment opportunities, traditionally on the second Monday in March. McGraw-Hill publishes *Advance Job Listings* (21) on a subscription basis, containing recruitment advertisements from 21 of their technical magazines.

ADVERTISING AND EMPLOYMENT AGENCIES

Of less interest to the student, but widely used by industrial employees seeking a new position, are employment advertisements and agencies. Trade magazines in most fields, and in particular *Chemical & Engineering News* (9), *Chemical Week* (22), *Science* (23), and *Chemical Engineering Progress* (10), carry columns of employment advertisements. In the society-sponsored magazines the rates to members, students, or unemployed members are quite low. *Chemical Engineering Progress* also carries a list of employment agencies or professional placement services. These groups conduct personal employment campaigns for a fee, which is sometimes paid by the new employer and sometimes by the job applicant. A volunteer group of this sort aiding engineers, scientists, and technicians was described in a recent article (24).

CONTACTING EMPLOYERS: LETTERS AND THE RÉSUMÉ

Once you have decided on the companies you wish to approach about getting a job, you must make contact with them. This is usually done either by letter or in an interview scheduled by the college placement office. Contact by telephone or unannounced visit is definitely not recommended: Lay the groundwork first, in writing, to provide a basis for discussion and allow time for the company to set up arrangements to deal with your inquiry.

In this section we consider the written contact only, since the interview (whether on campus or at the company location) is important enough to warrant separate treatment. We treat letters and the personal résumé separately, and further divide letters into several categories.

We want to emphasize the importance of communication skills in the selling job that these documents represent. There is no question about the need for proper grammar and spelling, clarity of statements, and all the other attributes of that ephemeral but essential quality called style, in the preparation of all this written material. In our experience, sad to say, it can represent a serious challenge to even the Ph.D. candidate to do this right. And a lot depends on it: You must rely on these letters and the résumé to introduce you to prospective employers, to attract and hold their attention, to arouse their interest in *you*, out of all the others vying for their consideration, and to convince them that you are the person to interview and, hopefully, ultimately hire. This is a classic sales job, and should be handled as such, though discreetly.

DIRECTED LETTERS

We mean here letters that are written to a few specific companies in which you are interested. The American Chemical Society (13) favors—and we agree—the approach in which you familiarize yourself with the company and its business, and mention this knowledge in a casual way in the letter; hence, it is directed specifically to that company. The ACS gives some examples of how the knowledge of the company can be worked in, but this should be fairly obvious. Just don't overdo it.

Both this type of letter and the undirected one described below should be brief and neat. Make them no longer than one page, and typed if at all possible. Be sure there are no obvious erasures, corrections, or smudges. Never send a carbon copy, but keep one for your records. The letter should contain a statement of the type of position you wish to apply for, in a general way (and here is where you can work in your knowledge of the company and the kind of business it does), and an indication of your qualifications for it. These can be done briefly, for the next step is to refer the reader to the enclosed résumé, which will amplify these statements with the necessary details. Then request the next step in the process, perhaps just the opportunity to discuss your qualifications with them, and close with thanks for their cooperation.

An important feature of both directed and undirected letters is the person to whom they are addressed. A directed letter should be sent to a specific person by name, if at all possible. This may be a contact you have developed or had given to you (with a request to forward your letter to, say, the director of personnel if your contact is a technical man), the personnel director (get his name by telephoning the company for that information, if necessary), or the director of research or other responsible management representative.

UNDIRECTED LETTERS

These letters, sometimes called "broadcast" letters, are those sent to a large number of prospective employers whose specific requirements are unknown or unimportant. Opinions differ as to how they should be formulated and how effective they are. They should be sent either to a responsible executive (company president or vice-president or director of research or engineering), or to the personnel office. Some say that you should attach your résumé; others suggest that personal information be included in the letter. Despite the fact that large numbers may be sent out (into the hundreds in bad times), these letters should be typed, not reproduced. We feel that the degree candidate looking for his first job should exhaust other possibilities before going to the undirected letter, but many experts feel that for changing jobs or answering ads, particularly in the engineering disciplines, this approach is very effective.

FOLLOW-UP LETTERS

After an initial contact has been made, follow-up letters are required for two reasons: One is to keep your name before the company while you are waiting for them to take action, and the other is to acknowledge some action on their part.

Keeping the company aware of your existence is usually of most value between the interview and the job offer (or rejection). You can write seeking additional information about the company, or simply asking where your application stands. Be sure to include thanks for their cooperation; and don't write too soon—normally it takes three to four weeks for the wheels to grind.

Except possibly for out-and-out negative replies to undirected letters, all communications from the company to you should be acknowledged promptly and courteously. This includes even a rejection; after all, you may want to approach the company again someday. Acknowledge the letter and thank them for considering your application. Reply promptly to a letter setting up an interview; drop a note of thanks to a campus interviewer after that meeting; acknowledge all offers, notifying the company of the date you expect to make your decision; decline other offers promptly as soon as you have accepted one; and in the letter of acceptance indicate when you will start work and other details that may be pertinent. In every case, express thanks, appreciation, and pleasure as appropriate.

THE RÉSUMÉ

After the prospective employer's interest has been sparked by your initial letter or the campus interviewer's report, he turns to your résumé to find out who you are, what you have done, what you know, and what you want to do. Here is the second opportunity for you to sell yourself, and you should plan this document to do just that in a concise, easy-to-read, dignified way.

Books have been written on résumé preparation (25); "kits" for preparing them are offered by employment agencies; examples are given in some of the sources we have cited (13, 14); a recent article (26) provides a valuable point of view; and your college placement office will have suggestions. Styles vary widely; consult a few and then, if you don't like any of them or feel your needs are specialized, don't hesitate to design your own.

In any case, your résumé should be limited to one page, possibly accompanied with a second page listing work experience, publications, or thesis titles: These are of secondary interest, but important enough to "add on" if you have them. It is advisable to have your résumé reproduced, preferably by lithography, in a large enough number to fill all your anticipated needs and then some. The cost is small, and a good-looking job results.

Table 10 is an example of a typical résumé, not as a recommendation, but to indicate what is important to include. Here are our comments, keyed to numbers on the form:

1. *Name.* Last name first seems to be the rule. The rest of the information required here is self-explanatory. Under foreign languages, indicate which are read, spoken, etc.
2. *Type of work desired.* This is one of the most critical parts of your résumé. Be sure that your career objectives are realistic in terms of your background. Consult the remainder of this book to familiarize yourself with the possibilities and the terminology to be used. Be rather

TABLE 10

Outline for a Short Résumé

RÉSUMÉ

Personal

Name (1) ____________________ Birth date _______ Soc. Sec. No. _________
Home address ____________________________________ Phone __________
College address __________________________________ Phone __________
Marital status ______ No. dependents _____ Citizenship _____ Height _____ Weight _____
Physical limitations _______________ Foreign languages _______________ Date available ________
Military status ____

Type of work desired (2)

____
____

Education

University	Location	Years attended	Degree and date	Major field

Thesis titles, publications (3) ____________________________________
____

Previous employment (4)

Dates	Employer	Description of Work

References (5)

Name	Title	Address	Telephone

specific, but not so much so as to limit your chances. If you have strong preferences as to location, list them here—but remember that too much restriction will narrow your opportunities and give an impression of immobility.

3. *Thesis titles, publications.* A separate list can be attached, but it is better to make this information brief. The older job candidates with several publications can state the number and general area, then list two or three of the most important ones.

4. *Previous employment.* A separate list can be attached. Put the information in reverse chronological order, latest job first. Again, limit the list to two or three items unless there are special circumstances.

5. *References.* Opinions are divided on whether to include names of references in the résumé. We feel that they should be included if the résumé is to accompany a directed letter or campus interview; otherwise, not. Only those who can discuss your technical qualifications should be listed. Be sure to ask permission before listing someone as a reference. Your college professors are usually willing to act as references—but be sure they remember exactly who you are!

THE CAMPUS INTERVIEW

This section is directed to the college student, graduate or undergraduate, who is about to make his first contacts with industry through the college recruiters. This is an important occasion. Salesmanship is required, and you should do what a good salesman would before a presentation: prepare and practice.

Begin by establishing and maintaining close liaison with your college placement office. Your major professor or department guidance counselor can advise you when to make this contact. The placement office will assist you in many ways, such as providing literature on the companies sending interviewers and advising on résumé preparation, in addition to scheduling the interview and providing the facilities for conducting it.

PREPARATION

Getting ready for the interview requires gathering two kinds of information: about yourself and about the company. At this stage it is imperative that you know not only what you want to do, but what you are best qualified to do. This will require some soul searching, analyzing your strengths and weaknesses, academic performance, aspirations, interests, and all the rest. This can't be done in an honest fashion over night; hopefully, you will have put a lot of thought on it before now.

The interview has a dual purpose. First, it gives you the chance to learn about the company and its opportunities in your areas of interest; and second, it allows them to learn about you. But it is short, typically 30 minutes, and there's no time for the exchange of trivial information. The interviewer will have studied your résumé; it's up to you to have studied information on his company, to gain some knowledge about its structure, products, policies, and plans. This knowledge is a must if you are to converse intelligently with the interviewer. The literature files of the placement office and the sources cited earlier in this chapter (14–17) should be consulted. You may also wish to look over lists of questions (27–29) you might ask the interviewer, if the answers aren't apparent in the material you have studied.

PRACTICE

If you are lucky, your placement office may have equipment (videotape, for example) and exercises for practice in the basic interviewing skills. If not, try to get a group of students together, including some who have had the experience, to try some mock interview sessions. You'll gain familiarity with what to expect and how to conduct yourself in actual interviews. Signing up strictly for practice is discouraged, but you might not wish to have your very first interview with the company you consider your best prospect.

ATTITUDE

Preparation for the interview includes getting your attitudes sorted out, too. Recruiters look for responsibility, maturity, a knowledge of where you're going. Be natural; don't try to make an impression. Dwell on the positive; the interviewer won't know your good points unless you bring them out. Keep alert; try to anticipate the questions the recruiter may ask you, and have sound answers ready.

CONDUCT

Your conduct at the interview starts with dress and appearance. This is an important occasion, so dress appropriately. College styles (including hair length and beards) are well known to the recruiter; you will have to decide for yourself whether they are in good taste at the interview and how important your appearance may be to the company.

Be on time for the interview, and don't try to prolong it when your time is up. Not only are these courtesies to the interviewer, who often has a full schedule, but they exemplify traits that employers find important.

Ask questions, at appropriate times, if you are puzzled about something or wish more information—but not because you think it is expected of you. Don't be persistent in asking about the nature of the job; it may indicate you haven't prepared for the interview. Leave matters of salary until later; it's too soon for the company to have evaluated your worth to them.

FOLLOW-UP

If any follow-up is required, take care of it promptly. Send whatever transcripts or other documents are wanted by the prospective employer as soon as possible. Get the recruiter's name, address, and telephone number so that you can call him if you need any more information.

THE PLANT TRIP

If the prospective employer is sufficiently interested in you, an invitation to visit his plant or research facility may be offered by letter a short time after the campus interview or other contact. This is a twofold opportunity: First, you have the chance to see his organization, meet a number of his employees, and find out what they do and what working and living conditions are at that location. And second, you may be sure that the company will be taking a close and careful look at you.

This last statement really sets the tone of the occasion: Salesmanship is again called for, since you must now impress favorably not only a campus interviewer but those who will hopefully soon be your bosses. This will be a concentrated dose, too, for you will be swept through a tight all-day schedule of appointments, tours, and interviews. In fact, it is likely that from the moment you step off the plane to the time you board it again on your return, you will be in the hands of (and under the scrutiny of) company personnel.

This is exactly as it should be, for the company is now considering an investment of hundreds of thousands of dollars in you and your career. Both they and you must be convinced that this is the correct choice. So act accordingly, both in selling yourself honestly and honorably, and in finding out as much as you can about the life you would lead as a company employee.

As to the mechanics of the trip, the preliminary arrangements should make it clear how you will travel, where you will stay, and about how much cash you will need to meet expenses. The amounts you pay will be reimbursed—more about this later—but be sure you come prepared. Some of these arrangements will be made in the letter of invitation; don't hesitate to ask about the others in your reply.

A large company will usually assign a relatively young employee, like yourself a few years hence, to be your host. He will supply transportation, escort you through the day's schedules, show you nearby residential and shopping areas, and take you to dinner. At the company, you will typically see employees doing the kind of work you are interested in, supervisors, and people in higher management. You will be asked many times about your background and training, your plans and aspirations, and your research experience. If you recently obtained your M.S. or Ph.D. degree, be prepared to talk about your thesis research, both informally and at a formal seminar. Even if you are not requested ahead of time to give a seminar, it is sometimes taken for granted that you can, and to be unprepared can be embarrassing. Bring slides if you have them already prepared; the company will usually have a projector available.

Although it is more likely that you will receive a job offer by letter after the trip, rather than at the end of the day, the matter of salary may come up. You will wish at least to have a general idea of what current salaries are in your profession, by having studied the annual reports on salaries published in the professional societies' news journals (9, 10).

The graduating student will probably find that the plant trip is the first instance of his contact with matters of professional ethics, a subject covered in more detail in Chap. 4. Two areas are pertinent here. First, even though it may not be mentioned specifically, it is appropriate to consider what you see and hear in a company's plant as privileged information. While it is unlikely that anything truly confidential will be disclosed to you, it is not appropriate to talk about a company's processes or technology, particularly to or at another company.

Second is the matter of the expense account whereby you and the company settle up the cost of your trip. You will be given forms to fill out, and guidelines as to what can be legitimately charged. Follow them scrupulously and honestly. Generally speaking, you may charge expenses essential to travel and living throughout the trip, including air, bus, train, and taxi fares, hotel or motel lodging, meals, tips, and the like. Entertainment expenses, and those which you would normally incur in any case, such as cost of haircuts and dry cleaning, should not be charged. Receipts must be attached for major expenditures, including air or train tickets and motel or hotel bills. Do not delay in completing and returning travel expense forms.

If you have more than one interview trip scheduled for a given area, it may be convenient to do them all at one time. In this case, the ethical procedure is to prorate the expenses among the various companies you visited. Never try to "make a buck" by duplicating or inflating expense accounts. It is not tolerated in employees or candidates.

Should there be a real mutual interest between you and the company, a second trip may be arranged. If you are married, this is an appropriate time to inquire whether the company will pay your spouse's expenses too, so that you both may look over the residential areas nearby.

Finally, as discussed earlier, don't neglect the courtesies of notes of thanks to the companies and, if appropriate, your hosts.

DECIDING AMONG COMPANIES

Let's assume that all has gone well in the previous stages and that you are fortunate to have one or more job offers, or at least several companies which you wish to compare with respect to career opportunities for you. How can you best evaluate these companies and come to the right decision among them?

This question is closely related to the one asked in the following chapter, namely how should the young professional conduct himself for the greatest mutual satisfaction with his company? Both deal with professionalism—the relationship built up between the professional scientist or engineer and his company, his discipline, and the scientific and non-scientific communities at large. Chapter 4 deals more extensively with many aspects of professionalism; here we will draw on one result of recent increased interest in the subject, a set of guidelines to professional employment (30) endorsed by a number of professional societies and included as Appendix I. This section is based in part on the guidelines for employers; and Chap. 4 on those for employees.

A recent article (31) provides a checklist for engineering job applicants based on the employer's guidelines. As it is being provided to senior engineering students by the National Society of Professional Engineers (32), we shall not repeat all of it. The checklist covers four aspects in which companies can be rated:

RECRUITMENT

This category includes not only recruiting technique (which tells a lot about the company and its attitude toward professional employees) but several other topics important in the early stages. For example, supervision: Which of the supervisors you saw in plant trips appeared the best qualified to help you learn the job and develop professionally? Which would you rather work for? How much authority over your salary and assignment does the supervisor have, as contrasted to those higher up? And manpower planning: Does the company do a good job in forecasting needs to provide for stability of employment? Finally, does it have a favorable employment contract, as described in Chap. 4?

TERMS OF EMPLOYMENT

Does the company appear to keep its employees well informed about its objectives, policies, and programs? How do its starting salaries compare with the going rates? What about performance reviews (more in Chap. 5), merit raises, and top salaries? How does it rate on fringe benefits—pension plan, health and life insurance, leaves? What is its attitude in general toward professionals: Do they have to join a union or punch a time clock, for example? How do the physical facilities and the availability of support staff (clerical, administrative, technical) stack up?

PROFESSIONAL DEVELOPMENT

What provision is made for you to continue your education, via tuition reimbursement, in-house courses, or leaves of absence for professional study? Is professional society membership and participation encouraged? Are technical publications encouraged? For the engineers, is the Professional Engineer registration encouraged and respected? In all these categories, inquire about and rate the company on actual participation rather than on written policy, since there may be a telling difference between the two.

MANAGEMENT

You will want to assure yourself that the company is well managed; some guidelines to aid you in this assessment are given in Chap. 11.

TERMINATION AND TRANSFER

You probably won't have asked about these policies—in fact, to do so turns some recruiters off—but they should be considered. Fortunately, they are usually covered in written material. Look into policies on termination pay, assistance in relocating, continuation of benefits, and reimbursement for expenses incurred in transferring from one location to another.

EQUAL EMPLOYMENT OPPORTUNITY

Though it is not a part of the checklist just discussed, this seems a good place to bring up the question of equal employment opportunity. Many responsible employers have long operated under a policy of equal opportunity for all in recruiting, hiring, training, on-the-job treatment, promotion, and benefit plans, regardless of race, religion, color, age, sex, or national origin. The spirit of such policies has now been supplemented by legal obligations incurred by all companies employing more than 50 people and holding more than $50,000 in government contracts—this covers virtually all but the very smallest industrial concerns.

These companies must have written "affirmative action" programs, and at each facility must meet goals for the proper utilization of women and minorities in job classifications within nine occupational categories, including professionals and technicians. Considerations of what proper utilization means are complex, including such factors as number of unemployed, percentage of minority and female work force, and availability of skilled personnel.

Management's responsibilities in implementing these requirements often go beyond what voluntary programs had earlier required. Policies in recruiting and hiring, training, and promotion are particularly affected. There is no question that the minority or female worker today finds far more nearly equal opportunities than ever before (33–36). Unfortunately, the requirements of proper utilization sometimes lead to what might be called reverse discrimination, in which a supervisor cannot get the person he honestly believes to be best suited or trained for a job because the company must attempt to meet a goal.

If the company you are interested in advertises as "An Equal Opportunity Employer," it will no doubt have available written material on its affirmative action compliance program.

SOME ADVICE TO STUDENTS

One of an author's problems is how to get his message to the proper audience. At this point we are acutely aware of the difficulty, since we want to advise the undergraduate student, in his early years, on how to choose his curriculum and college activities to best fit him for an industrial career. We suspect that for most of our readers this advice will come too late. But if you are in this category, perhaps you will serve as our proxies to carry the message of this section to your younger colleagues. Tell them to

- Become informed about industry and what goes on in it by reading books such as this one and others (37–39), articles (40), and journals in their field such as *Chemistry* (41), *Chemical & Engineering News* (9), *Chemical Engineering* (42), *Chemical Engineering Progress* (10), and *Chemical Technology* (43).
- Take elective courses in the humanities and social sciences which may be valuable later, such as economics, business administration, marketing, or management.
- Take advantage of other relevant college programs, such as professional orientation courses, cooperative work-study programs with industry, team study projects, and technical curriculum electives in other fields to develop an interdisciplinary approach.
- Join a professional society—most have inexpensive student affiliate memberships—and take advantage of plant trips, talks by industrial employees, and other programs.
- Talk to professors, friends, and associates about their experiences in and with industry.
- Develop a positive attitude toward industry early by getting to know what it is all about.
- Finally, review the first section of this chapter on what industry looks for in new employees, and cultivate these characteristics throughout college and graduate school.

SOME ADVICE TO NEW EMPLOYEES

The fateful decision has been made, the job offer has been accepted, old ties have been severed, and you are now a young professional in the industrial community. What can we say to wish you on your way?

First, you are just beginning and there is still much to learn—if this were not so, our book would be ending here. You will have many entirely new situations to face. You are now a competitor in the often cutthroat world of business. And with the benefits of industrial employment come many responsibilities, some of which make up the subject of Chap. 4.

The trials and tribulations of existence in the business world have been the subject of several best-selling books (44–46) in recent years; a brief look at them should be amusing, enlightening, and informative at this stage in your career. Finally, here are a few of the new situations you may be faced with and should watch out for:

You will be called on to relate to nonprofessional people. They can be very perceptive of the way you feel toward them. Recognize their value and earn their respect.

There will be intervals in your career when your capacity is not fully utilized—for example, right after you are hired, before you have learned enough to be of full value to the company. Recognize that these periods are temporary, and utilize them for self-improvement.

A busy boss can be very difficult to communicate with. You will have to look carefully for the proper avenues.

You may find your supervisors reluctant, in performance reviews, to level with you about your personal weaknesses and preferences; it is easier for them to put the emphasis on your strong points. Yet you must learn where improvement is required in order to progress.

The "way things are done" may seem to be based on nothing more than abysmal stupidity. This probably means that you don't have all the inputs needed for understanding the situation.

Be prepared at all times to fall back on alternative courses of action—preferably more than one so things don't get polarized. The one that seems "far out" may be the best.

Learn the value of time, and of reducing the time lapse between setting an objective and achieving it.

CHANGING JOBS

If you are reading this chapter in preparation for seeking a different, rather than a first job, don't feel that your status is unusual. About 10% turnover per year is customary in the chemical industry, and the average stay of a person in his first job is said to be three years. Though surprisingly short, this seems about right: Some time ago one of us (47) suggested two to five years as a minimum goal that would provide an adequate "first experience" for the employee and compensate the employer for his not inconsiderable replacement cost. It was recently estimated (48) that to replace one man costs $10,000 for recruiting, relocation, orientation training, and loss of research progress in the interim.

Much of what has been said in this chapter is applicable to one who is planning to change jobs, but there are some significant differences. Before considering them, however, we suggest you consider carefully your motives for a change. Is it really called for? Presumably—unless you have had the misfortune of being fired—the decision has stemmed from feelings of disssatisfaction. Would it not be wise to attempt to analyze them, to consider areas of satisfaction as well, and to rate your present job as a whole with respect to the degree of satisfaction you feel? Then see how many of the areas rated low can be resolved by changes in attitude or other means short of quitting. Can you be sure of doing better next time?

A rating chart for job satisfaction and lists of employees' most commonly cited reasons for satisfaction or dissatisfaction, based on a survey of nonsupervisory scientists and engineers, was published a few years ago (49). Rather than reproduce the results in their entirety, we excerpt some of the more important conclusions.

Reasons most often cited as causes of dissatisfaction were, in order of citation frequency:

1. Required to perform routine or continuous tasks, such as calculating or plotting.
2. Limited by procedures or organization of higher management.
3. Required to prepare and present reports.
4. Required to perform clerical tasks.
5. Inability to conduct research, experiments, or tests successfully.

It strikes us that several of these categories carry the implication that self-improvement would alleviate the problems.

On the other side of the coin, reasons for satisfaction most frequently cited were:

1. New, nonroutine, challenging work.
2. Personal pride in the work.
3. Realization of personal goals, through utilization of capabilities and training.
4. Visible results, and completion of tasks.
5. Relations with and recognition from people.

Rate your job numerically according to Table 11 before making a final decision.

If your decision to change is still unshaken, then you have some more serious thinking to do.

You are now more experienced than the recent graduate looking for his first job. You have more to offer an employer, and he will expect more from you. To sell yourself, you will have to resharpen your communication skills, including preparing a new résumé. Now you must be prepared to know and state clearly what you can really offer an employer. This will probably be quite different from what you thought as a student. Some recommend "sensitivity training group" participation as a means of helping you think these things out.

At this time you should also consider the value of additional education, in the form of refresher courses or the like, to broaden or intensify your training in specific areas.

TABLE 11

Rating of Satisfaction with Conditions of Present Job*

Below is a list of the major aspects of your job. Check in the appropriate column the degree of satisfaction you feel with regard to each of the aspects. Try to rate each aspect as independently of the others as possible.

	Very satisfied	Fairly well satisfied	Neutral, mixed feelings	Somewhat dissatisfied	Very dissatisfied
Job activities (overall rating)	______	______	______	______	______
Facilities for your work (equipment, laboratory, apparatus, library, etc.)	______	______	______	______	______
Hours of work (starting and closing time, overtime, etc.)	______	______	______	______	______
Assistance furnished by technicians, clerks, etc.	______	______	______	______	______
Working relations: (a) as a team member working with others on assigned problems	______	______	______	______	______
(b) with other professionals with whose specialities your own work has to be coordinated	______	______	______	______	______
(c) with nonprofessionals, such as operating executives in production, and purchasing	______	______	______	______	______
Nonworking relations with others in the company (at lunch time, en route to and from work, after working hours, etc.)	______	______	______	______	______
Actions and attitudes of immediate supervisor	______	______	______	______	______
Salary, salary adjustments	______	______	______	______	______
Benefit plans (insurance, pensions, etc.)	______	______	______	______	______

TABLE 11 (continued)

Intangible rewards (recognition privileges, challenge of job, seeing your ideas carried out in practice, etc.)	____	____	____	____	____
Information received regarding company matters	____	____	____	____	____
Progress and present placement in the organization	____	____	____	____	____
Company assistance to professional development	____	____	____	____	____
Relations with members of top management	____	____	____	____	____
Top management's administration of professional work in the company (general policies, etc.)	____	____	____	____	____

* Adapted from (50). Reprinted by permission of the University of Michigan.

Then set about making contact with prospective employers. As noted earlier in this chapter, most seekers of second and subsequent jobs tend to rely more on advertising and employment agencies than do graduates, but the other routes should not be overlooked. Above all, proceed with care: As a professional, you have a lot at stake when you change jobs.

You may wish to read the results of a MIT symposium on changing careers (50). If you are considering a change as a result of company acquisition or merger, a recent article (51) may be helpful.

REFERENCES

1. "Correlating Chemical Education with Industry Needs," Chap. 24 in *Chemistry and the Economy,* American Chemical Society, Washington, D.C., 1973.
2. B. W. Rossiter, "What an Industrial Laboratory Desires in the Preparation of Science Graduates," *J. Chem. Educ.* **49**, 388–391 (1972).
3. George S. Odiorne and Arthur S. Hann, "Effective College Recruiting," Report No. 13, Bureau of Industrial Relations, University of Michigan, Ann Arbor, 1961.
4. Fred W. Billmeyer, Jr., *Textbook of Polymer Science,* 2nd ed., Wiley-Interscience, New York, 1971.
5. B. R. Schlenker, *Introduction to Materials Science,* Wiley, New York, 1969.

6. J. M. Tedder, A. Nechvatal and A H. Jubb, *Basic Organic Chemistry, Part 5: Industrial Products,* Wiley, New York, 1975
7. T. R. Dickson, *The Computer and Chemistry,* Freeman, San Francisco, 1968.
8. Thomas J. Hailstones, *Basic Economics,* 3rd ed., Southwestern, Cincinnati, 1968.
9. *Chemical & Engineering News,* published by the American Chemical Society (11).
10. *Chemical Engineering Progress,* published by the American Institute of Chemical Engineers (12).
11. American Chemical Society, 1155 16th St., N.W., Washington, D.C. 20036.
12. American Institute of Chemical Engineers, 345 East 45th St., New York, N.Y. 10017.
13. ACS Committee on Professional Relations, "Finding Employment in the Chemical Profession," American Chemical Society (11).
14. *College Placement Annual,* College Placement Council, Bethlehem, Pa.
15. *Chemical Guide to the United States,* Noyes Development, 4th ed., 1966.
16. *Industrial Research Laboratories of the United States,* Bowker, New York, published annually.
17. *Thomas' Register of Manufacturers,* Thomas, New York, published annually.
18. Anon., "Career Planning and Directory '75," *Chem. & Eng. News* **52** (40), 41–47 (Oct. 7, 1974).
19. *Business World,* Placement Publications, Inc., Box 1234, Rahway, N.J. 07065.
20. *New Engineer,* MBA Communications, Inc., 555 Madison Ave., New York, N.Y. 10022.
21. *Advance Job Listings,* McGraw-Hill Publications Co., P.O. Box 900, New York, N.Y. 10020.
22. *Chemical Week,* P.O. Box 386, Times Square Station, New York, N.Y. 10036.
23. *Science,* American Association for the Advancement of Science, 1515 Massachusetts Ave., N.W., Washington, D.C. 20005.
24. Fredric A. Litt, "Unemployed? Try VEST for Size," *New Eng.* **2** (6), 6–7 (1973).
25. M. P. Jaquish, *Personal Résumé Preparation,* Wiley, New York, 1968.
26. Guy Weismantel and Jay Matley, "Is Your Résumé Junk Mail?" *Chem. Eng.* **81** (24), 164–168 (Nov. 11, 1974).
27. Jack G. Calvert, James N. Pitts, Jr., and George H. Dorion, *Graduate School in the Sciences—Entrance, Survival and Careers,* Wiley-Interscience, New York, 1972.
28. Dimitrios Tassios, "The Interviewing Process," *Chem. Eng. Prog.* **69** (7), 26–28 (1973).
29. The Editors of *Chemical Engineering, Secrets of Successful Job Hunting,* McGraw-Hill Publications, New York, 1972.
30. "Guidelines to Professional Employment for Engineers and Scientists," available from the AIChE (12) as well as other endorsing societies.
31. Gayle N. Wright, "An Employment Guidelines Checklist for the Engineer Job Applicant," *Prof. Eng.* **43** (8), 40–43 (1973).
32. National Society of Professional Engineers, 2029 K St., N.W., Washington, D.C. 20006.
33 Herbert Popper, "Creating New Educational and Job Opportunities," *Chem. Eng.* **79** (21), 159–164 (Sept. 18, 1972).

34. Adolph Y. Wilburn, "Careers in Science and Engineering for Black Americans," *Science* **184**, 1148–1154 (1974).
35. Jane S. Shaw, "Is Your Male Chauvinism Showing?" *Chem. Eng.* **80** (18), 98, 100 (Aug. 6, 1973).
36. C. F. Fretz and Joanne Hayman, "Progress for Women—Men Are Still More Equal," *Harvard Bus. Rev.* **51** (5), 133–142 (Sept.–Oct. 1973).
37. J. H. Saunders, *Careers in Industrial Research and Development,* Dekker, New York, 1974.
38. David Allison, ed., *The R&D Game,* MIT Press, Cambridge, 1969.
39. Guy Alexander, *Silica and Me—The Career of an Industrial Chemist,* American Chemical Society (11), Washington, D.C., 1973.
40. Guy Alexander, "A Chemist in Industry," *Chemistry* **45** (1), 19–21 (1972).
41. *Chemistry,* published by the American Chemical Society (11).
42. *Chemical Engineering,* McGraw-Hill Publications Co., P.O. Box 900, New York, N.Y. 10020.
43. *Chemical Technology,* (*Chemtech*), published by the American Chemical Society (11).
44. Lawrence J. Peter and Raymond Hull, *The Peter Principle,* Morrow, New York, 1969; Bantam Books, New York, 1970.
45. Robert C. Townsend, *Up the Organization,* Knopf, Westminster, Md., 1970; Fawcett, New York, 1971.
46. Shepherd Mead, *How to Succeed in Business Without Really Trying,* Simon & Schuster, 1952.
47. Fred W. Billmeyer, Jr., "Careers in Plastics: College Teaching," *SPE J.* **23** (3), 33–34 (1967).
48. Sherman Tingey and Gordon C. Inskeep, "Professional Turnover," *Chemtech* **4**, 651–655 (1974).
49. Lee E. Danielson, "Characteristics of Engineers and Scientists," University of Michigan, Ann Arbor, 1961.
50. Sanborn C. Brown, ed., *Changing Careers in Science and Engineering,* MIT Press, Cambridge, 1972.
51. Walter J. Bray, "Engineer in an Acquisition," *Chem. Eng.* **80** (2), 116–122 (Jan. 22, 1973).

CHAPTER 4

PROFESSIONAL RESPONSIBILITIES

When you become a new employee, you stand at the beginning of a successful industrial career. But you will soon discover that you have acquired new responsibilities, and associated with these are many ethical, legal, and moral questions which you probably have not had to consider before. Although success in your career depends on your technical ability, equally important is how you live up to the responsibilities you must shoulder as a professional, and whether you accept them with an enthusiastic and positive approach. These factors require periodic reappraisal, so even a seasoned employee benefits from a revitalizing look at how he's meeting his responsibilities.

A moment ago we described you as a professional, and we should clarify that term. Most of us think of doctors and lawyers as professionals, but how about chemists and chemical engineers? Well, the Supreme Court recognized chemistry as a profession before the turn of the century, and the National Labor Relations Act of 1947 (the Taft-Hartley Act) described a professional as one whose occupation is predominantly intellectual and nonroutine, requires the consistent exercise of judgment and discretion, cannot be characterized by a standard output per unit time, and requires specialized study and advanced knowledge customarily obtained in an institution of higher learning.

Finally, the American Institute of Chemical Engineers defines (1) a professional as "an individual who, with adequate training, experience, intellectual capacity, and moral integrity, effectively devotes his skills and knowledge to the service of society and his profession in whatever assignment he finds himself, being fully sensible of the personal responsibility and trusteeship conferred by his special training."

Thus by the accepted definitions you are indeed a professional. As such you are expected to accept certain responsibilities and to meet certain standards significantly beyond those of a purely technical nature. These legal, moral, and ethical responsibilities fall into four major categories, to which we devote this chapter: professional ethics, employer–employee relations, maintaining technical competence, and shared responsibilities.

When you first leave the university and accept a job, you must be concerned with what industry expects of you as a professional, aside from technical matters; this involves *professional ethics.* Once you are established in industry, you find the need of developing mutually acceptable relations with your employer and of conforming to the existing standards for *employer–employee relations.* As time passes, you will see the necessity of maintaining your technical capabilities at a high level of competence and avoiding obsolescence—*sharpening your skills* (don't laugh—people can become obsolete even more quickly than machinery). And from the beginning to the end of your career you will find that you must meet responsibilities to your family, your professional colleagues, your community, and the world at large (*shared responsibilities*) as well as to yourself, your employer, and your profession. In the following sections we examine these four areas of responsibility.

PROFESSIONAL ETHICS

THE EMPLOYMENT CONTRACT

If your employer is typical, one of the very first requirements of the new job is to formalize your major responsibility to your employer by signing an employment contract or agreement. Most employers require this as a condition of employment. It usually includes statements that give to the company exclusive ownership of patents developed at their expense, and that prohibit the disclosure of any company confidential information which the employee may have acquired.

A typical employment agreement (2) is shown in Appendix II. Its relatively simple clauses cover the two areas mentioned above and provide for the return of all confidential information when employment terminates. The concern over patents is justified, for they provide the major mechanism by which the employer can protect his investment in research and development. In the United States and many other countries, patents are granted only to individual inventors and must subsequently be assigned by the individual to his company. Patents and related matters are discussed more fully in Chap. 10.

Some employment agreements specify further obligations of the employee. It is often agreed that he will not, while employed, enter into any additional jobs or activities that would conflict with his employment or impair his performance. More controversial are the postemployment contracts required by some employers, which preclude the employee from working for competitors for a specified time after termination of the job covered by the contract. Such an agreement is needed only when the employee will be involved in an unusually important confidential development.

It is very important that you fully understand the intent and contents of the employment agreement *before* you sign it. (The legality of these contracts is upheld by the courts.) Most of them are fair and benefit you as well as your employer. The ACS (3), the AIChE (4), and other professional societies take an active role in explaining employment agreements to their members and have brochures on them available. For a fresh view, see ref. 5.

TRADE SECRETS

After you have signed your employment agreement and fulfilled all other conditions of employment (including a medical examination), you and your employer will begin to explore the new responsibilities each of you has undertaken. Your employer will inform you of the many policies of the corporation of which you have become a part. Here we wish to discuss those policies dealing with an especially important area, that of confidential information you will receive, generally known as "trade secrets." What are they? How are they protected (in addition to clauses in the employment contract)? And how will knowledge of them affect you?

Legally, a trade secret may be any formula, pattern, device, or compilation of information which is advantageous to the business and not available to a competitor. It may be a chemical formula, a process for manufacturing or formulating, a blueprint for a machine, or a list of customers. In deciding what is a trade secret, one should consider the cost to the company (in money and effort) to get the information, the amount and duration of the competitive advantage it provides, the difficulty a competitor would have in getting the information through his own research efforts, the extent to which the secret is known to employees, and what protective measures are taken to safeguard its secrecy (6).

Some examples of trade secrets are customer lists, discount schedules, marketing plans, or purchasing specifications; specific formulas and formulation procedures; design drawings; analytical test methods developed within the company; information on production costs, process rates and the like; patentable ideas, patent applications and preliminary reports on inventions; lists of research and development programs under way;

figures on production capacity, waste, manpower requirements, sales, costs, or profits; computer programs; and lists of suppliers and quantities of materials purchased.

A trade secret has been compared to a bottle of perfume: If it is permitted to leak out, its value to the owner is gone (6).

How are trade secrets protected? Here are some common practices which companies use:

- Employment contracts.
- Precautions to avoid conflict of interest.
- Release of information only on a "need-to-know" basis where it is not possible to get patent protection, (Generally, companies prefer to seek patent protection where possible—see Chap. 10.)
- Internal communications on secrecy, reminding employees of their obligations to protect company secrets at all times. Figure 14 is an example, pointing out the danger of casual employee conversations in public places.
- Termination interviews to remind the departing employees of their obligations, which in some cases include restrictions against working for competitors in the same area for specified periods of time.
- Precautions to ensure that visitors do not gain access to company secret information.

FIGURE 14. A typical poster on confidentiality (2, from the *3M Megaphone*). (Reprinted by permission of The Conference Board.)

How is the technical man affected by knowledge of trade secrets? First of all, he needs to learn to distinguish between what is a trade secret and what is not. Unfortunately in some cases it is virtually impossible to distinguish between employee know-how and company confidential material. When the employee changes jobs, confusion on this point can lead to real trouble, as evidence by several widely publicized court cases. A number of articles and books review this subject (6–12).

Some constructive suggestions cited by the ACS on how to handle trade secrets, for the employer as well as the employee, follow*:

For the employer,

- Use an employment contract containing a disclosure clause.
- When hiring someone previously employed in industry, consult the former employer about any job restrictions.
- Ask the new employee about any such restrictions or ethical problems.
- Educate, and periodically remind, employees about trade secrets and their importance.
- Don't place an employee on a compromising job.
- Limit important secret information on a need-to-know basis, and clearly identify secrets as such.

Suggestions to the employee include

- Adhere to high standards of personal and professional ethics, respecting and guarding the concerns of your employer like your own.
- Read and understand your employment agreement.
- If you change jobs, tell your new employer what kind of work you were doing, and make it clear you will not disclose confidential material to him.
- Distinguish carefully between what is generally known and the confidential material you have learned on the job.
- Consider whether it is worth the risk and effort to take a job where you will have to withhold trade secrets that may be valuable in performing your new duties.
- Do not take any written records, samples, notes, etc. (even your own) when leaving a job, without specific authorization.
- Exercise due care when discussing company work in public, such as at technical meetings.
- If you do become involved in a trade secret dispute, get a lawyer who is familiar with the subject!

* From the ACS 1968 booklet *Trade Secrets . . . Ethics and Law*, American Chemical Society, Washington, D.C., 1968. Reprinted by permission of the American Chemical Society.

Unfortunately, the greatest threat to the integrity of a company's confidential information is its employees. Most often they contribute to its loss through carelessness rather than deliberately. Remember Fig. 14!

THE RIGHT TO PRACTICE ONE'S PROFESSION

It is clear from the foregoing that one critical period with respect to trade secrets arises when an employee plans to change jobs. Most companies rely on his professional ethics and honesty to avoid disclosure of company confidential information, but frequently he is asked to sign an agreement not to work for competitors for a specific period of time. In many cases, he is provided compensation if he cannot get another job because of his specialized training. Many companies will not hire an applicant for work in a sensitive area identical to that in which he was previously employed.

Still, the question remains whether an employee familiar with trade secrets can ethically and honestly work for a competitor within his specialized field. Let us examine this question by reference to a well-publicized example, the case of Donald Hirsch.

The Hirsch Case

Dr. Hirsch, an employee of the Du Pont Company for twelve years, had spent six years on research and development of a new chloride process for making TiO_2. Increasingly dissatisfied with the lack of job openings in management, he began to look for new job opportunities. He applied to some ten companies over a three-year period, but without success.

In 1962, the American Potash and Chemical Corporation (Ampot) began recruiting personnel for a new TiO_2 plant to use a chloride process developed by a third company, taking full precautions to prevent disclosure of confidential information by job applicants. Hirsch applied and, concluding that Ampot wanted his skills and general knowledge rather than Du Pont's proprietary knowledge, accepted the position of manager of technical services for the new plant then being planned. Fifteen years earlier, he had signed an employment agreement with Du Pont containing the usual disclosure clause. As a condition for employment at Ampot he had signed an agreement with them that he would neither disclose to nor employ for Ampot any confidential information or trade secrets learned on his earlier job.

Du Pont immediately obtained a court restraining order preventing Hirsch from working on the TiO_2 project at Ampot, contending that it would be impossible for him to do so without divulging or employing the confidential knowledge received at Du Pont. Ampot was forced to assign Hirsch to a position elsewhere. The case came to trial two years later, and was finally dismissed some three years after that. The legal arguments were far too involved to review here, but it is clear that the court was fully aware of the conflict between the protection of trade secrets and the protection of the individual's right to use his knowledge and skills in gainful employment.

Although each case is likely to be different, it is clear that the best practice is to avoid changing jobs within the same specific area in which you know trade secrets. Similarly, most companies take extensive precautions against hiring someone into such a position.

INDUSTRIAL ESPIONAGE

To most of us, the concept of espionage exists only in the television spy movie, and many would certainly adopt an "it can't happen here" attitude toward espionage in relation to industrial trade secrets. Yet it does happen, to the tune of losses estimated at $4 billion annually! As professionals, we all have the responsibility of guarding against loss by espionage of the trade secrets we share. In addition to the precautions mentioned earlier and a recent book on the subject (7), here are a few suggestions:

- Always see that confidential material is locked up when you are out of the office, even over the lunch hour. Don't overlook bulky items such as presentation charts as well as documents and notebooks.
- Never discard confidential information into a waste basket. Most companies have special locked disposal containers or paper shredders.
- Limit the distribution of confidential reports to a "need-to-know" basis.
- Never take a complacent approach toward the security of confidential information.
- Always report promptly, to your supervisor or security group, any suspicious or unusual activities, both at work and outside.

You may find, particularly as a newcomer, that you are on the receiving end of some of these precautions; don't let it bother you. In most companies, only those who have been employed for several years and have had the chance to develop and demonstrate their loyalty and integrity participate in highly confidential projects.

The Aries Case

Occasionally a spectacular espionage case brings home the fact that "it can happen here." One such involved Dr. Robert S. Aries, a chemical engineer. A naturalized citizen, educated in the United States, Aries was highly respected for his technical articles and patents. But he took every opportunity to "pump" his associates for proprietary information, and sold these pirated trade secrets.

This espionage was uncovered by chance. As Merck and Company was about to market a new anticoccidial agent for destroying poultry parasites, Aries presented a technical paper in Canada describing a similar drug he had developed. Considering the time and effort Merck had invested, it seemed highly unlikely that Aries could have developed such an agent independently. By chance, Merck was then completing negotiations to purchase a French chemical company, whose management reported to Merck that Aries had licensed them to manufacture and market his agent. Further investigation showed that he had also licensed companies in the United States, Great Britain, and Switzerland, for sizable sums of money. The formulas for the two drugs were identical.

It was discovered that a former student of Aries, employed by Merck but not on the same project, had stolen and copied documents and provided them, together with samples, to Aries. In some cases Aries transmitted the documents to his licencees without recopying them, so the trail was easy to follow. Subsequently it was found that secrets regarding a lubricating oil additive and electrical components for computers had been obtained by Aries from other former students.

Prosecution followed, and judgments of more than $21 million were awarded to Merck and the other two companies whose secrets had been pirated. But Aries had fled to Europe and his American companies were insolvent. Despite civil and criminal actions pending against him here and abroad, he has not been brought to trial.

CODES OF ETHICAL CONDUCT

Our entire way of life is based on ethical concepts which have evolved over the centuries. Part of the responsibilities of a profession and a professional individual is adherence to sets of principles of professional ethics. Ideally, the mutual benefits of an open and honest relationship between employer and employee ought to be obvious. But in the real world, with all its complexities, this isn't always so, and the professional employee must constantly review his motivations, off the job as well as on. Unfortunately, the difference between right and wrong may not be clearcut, and decisions must be based on previous experience and personal values. References 13 and 14 discuss the problems of professional ethics in a general way.

Three major professional societies relating to chemistry have published codes of ethics. The ACS Chemist's Creed (15), which is short enough to reproduce in its entirety in Appendix III, is the most recently adopted of the three. The Canon of Ethics of the AIChE (16) is divided into four sections: fundamental principles of professional engineering ethics, relations with the public, relations with employers and clients, and relations with engineers. Several specific standards of conduct are set forth in each section. The Code of Ethics of the American Institute of Chemists

(17) includes 21 detailed rules of professional conduct and ethics, similar to those advanced by the ACS and AIChE.

WHAT WOULD YOU DO?

Many of these ethical principles are very general. How would you apply them in day-to-day situations? Here are some typical problems, but you will have to supply the answers, weighing the pressure to succeed and the temptation to compromise your standards. We hope you will conclude that you only cheat yourself by succumbing to temptation.

- You are traveling on a business trip, and overhear two nearby passengers, from a competitive company, discussing confidential marketing plans for a new product. Should you pass the information on to your supervisor when you return? Or should you identify yourself to these people rather than eavesdrop?
- You had an idea a few months ago and mentioned it to an associate. He pursued it, and is filing a patent application without giving you credit. Do you tell anyone, and if so, who? What if you had a written record of your conversation, showing the idea was originally yours?
- At Christmas, time, several suppliers with whom you deal send you gift bottles of liquor. One, who has just entered a sizable bid for a purchase contract, sends you an expensive set of golf clubs. Should you accept these gifts?
- You discover a hazardous situation in the production operation you are working on. Your supervisor tells you that it has been known for some time, but they can't shut down to correct it because the product is critically needed. Do you go over his head and report this to higher management? Is it right to continue production under such circumstances?

In reviewing these questions, remember that someone else's unethical conduct should never influence your decision. "A person must learn to think in three dimensions: as an individual with responsibilities to himself, as an employee with loyalties to his company, and as a citizen with a duty to the well-being of society" (13).

EMPLOYER–EMPLOYEE RELATIONS

THE UPSURGE OF PROFESSIONALISM

The early 1970s were difficult times for many professionals because of widespread layoffs in the chemical and some other industries. Unemployment was high and job offers almost nonexistent. As a consequence, there was an upsurge of interest in the problems of the professional and their solutions. Professionals were dissatisfied with the practices of some employers (18, 19) in areas such as layoffs, vested pension plans, handling of terminations, salary compression, lack of communication with management, the status of the professional and his professional society, and indications that the company placed little value on professionalism.

Many scientists and engineers have turned to their professional societies for help in solving these problems through the development of employment policies and practices that are fair to both employer and employee. Many employers have joined in these programs with sincere interest.

The area of employer–employee relations is a very controversial one; emotions frequently get very hot! The range of employee attitudes is wide, from the activists who demand immediate change for change's sake (if it satisfies their goals) to those who have no desire to get involved. As is so often the case, a middle course seems advisable. We feel that one should be open-minded, not let emotions or prejudices obscure judgment, get all points of view on the subject, and get involved by working through the proper channels within the company to effect any policy changes that may be needed. Among these channels are new legislation protecting professionals, increased activity by professional societies, and increased activity by unions.

There are two sides to every question. The side of the professionals who feel mistreated is dramatized in a recent book (20) which is quite critical of the treatment of engineers and scientists in industry. It makes the point that these people are professionals and should be treated as such. The side of the employers must not be overlooked, and the outstanding point here is that most companies—the vast majority in our experience—do provide an excellent working environment, recognize the needs, motivations, and aspirations of their professional employees, and abide by policies and practices that treat them as professionals.

PROFESSIONAL SOCIETY GUIDELINES

Very significant progress in employer–employee relations was made in 1972–1973 with the development and widespread acceptance (21) of sets of guidelines through cooperative action of professionals, societies, and companies. One of these, the Guidelines to Professional Employment for Engineers and Scientists (Unified Guidelines) (22), was prepared by 16 major professional societies, and in less than a year had been endorsed by 22 societies with a combined membership of over half a million professionals. We feel these guidelines are important enough for you to have at hand, and have included them as Appendix I. A similar set produced by the ACS (23) differs somewhat in the treatment of vesting and termination conditions. A number of recent articles discuss guidelines in general (24) and those of the ACS in particular (25, 26).

The intent of these guidelines is to provide checklists of important factors for establishing a good climate for both the employer and his professionals. They are guidelines for self-analysis, and for both sides they represent desirable goals rather than specific minimum standards. The Foreword (page 262) summarizes their purposes nicely.

Most companies meet the goals of these guidelines well. A number of major companies have formally endorsed the Unified Guidelines; among the first were the Dow Chemical Co. and Dow Corning (27). In a letter to all salaried employees, Dow Chemical compared company policies and practices to the guidelines, point by point. They corresponded very closely, with some differences only in pension policies.

MOONLIGHTING

If you maintain regular outside employment in addition to your industrial job, you are moonlighting. This is not at all unusual, and most companies—80% in a recent survey (28)—neither expressly permit nor forbid it. Many do place restrictions on the type of outside employment, however, in order to avoid conflicts of interest or deterioration of job performance. An outside job should not affect your ability to carry out company assignments, including working overtime when required. Employment by competitors, vendors, or customers is not only unethical but also represents a conflict of interest and is almost always prohibited. Clearly, any employee who has access to confidential information should not undertake any outside employment in which he risks jeopardizing the security of that information, knowingly or inadvertently.

If you are considering an outside job, the best policy is to discuss it openly with your supervisor before you accept it. Your ambition certainly won't be held against you, and possible later embarrassment will be avoided.

EXPENSE ACCOUNTS

Your first experience with a company travel expense account was no doubt at the time of your interview trip, as described in Chap. 3. On the job, you will first be reimbursed for your expenses in moving to the job site, and many times more as you take business trips at company expense.

Most companies state their expense accounting policies in a set of instructions accompanying the travel expense report form. Records of such business expenses are required by law, so procedures don't differ much among companies and we don't need to describe them in detail. Here are a few guidelines:

- Include only reasonable expenses incurred as a result of company-authorized business. That means you must exercise discretion in what you charge; don't abuse the privilege.
- It may be convenient to use your personal automobile for transportation, in which case you are reimbursed at a company-determined fixed rate per mile plus turnpike tolls and parking fees. Get permission before using your car on a trip of several hundred miles. If the cost of using your car is considerably greater than that of commercial transportation, your reimbursement may be limited to the commercial amount.
- Obtain and include with your report receipts for lodging and commercial transportation (certainly plane and train; company policies differ for bus and taxi). Receipts may also be needed for meals or for large or unusual expenses. Entertainment costs require prior permission and full documentation; again, check company policies.
- If you expect to spend a considerable amount of money on a company trip, don't hesitate to ask for a cash advance. Many companies also purchase the tickets for commercial transportation for you.
- Especially if you have an unspent balance from an advance, but as a matter of good relations in any case, complete your expense statement as soon as possible on your return. And don't forget that it is highly unethical to "pad" the figures.

DRUG ABUSE AND ALCOHOLISM

These two related personal problems affect employee behavior and productivity by causing absenteeism, inefficiency, and job turnover in addition to decreased morale (29–31). And statistics show that, although it may seem highly unlikely, you could develop either of these problems; they are not limited to particular social or economic levels. It is estimated that about 3% of the working population has an alcohol problem, and in a recent survey (30) about ⅔ of the companies responding viewed drug abuse as a current or incipient problem to them.

The real difficulty with alcohol appears to be the fact that it is socially accepted in moderate amounts, as we are all aware. The thin dividing line between use and abuse is all too easy to cross, as demonstrated by the terrible cost of drunken driving as well as addictive alcoholism. It would seem that any use of alcohol carries moral and ethical responsibilities too often overlooked.

To the over-30 generation, the use of drugs is not at all socially acceptable, but "soft" drugs are becoming increasingly available and gaining social acceptance by young people. The moral and ethical problems involving soft drugs would seem to be basically similar to those caused by abuse of alcohol.

Our purpose, however, is to comment on the attitudes and policies of most companies on these problems, rather than the attendant social and moral questions. As you would expect, the possession, sale, purchase, transfer, or use of any alcoholic beverages or drugs (except those prescribed for medical care under a doctor's supervision) is strictly prohibited on company premises. Violation is serious, and the employee may be severely reprimanded or dismissed. If an employee comes to work obviously under the influence of alcohol or drugs, he will probably be refused entry by the security guard at the gate. If he does get in, his supervisor should send him to the medical department. Such precautions are particularly important in the chemical industry because of the hazardous nature of many of the operations.

Despite widespread concern over their seriousness, many companies prefer to treat the problems of alcoholism and drug abuse by considering each case individually rather than setting fixed policies and practices. Action may range from referral to medical treatment and counseling, to discharge and notification of the police if the law has been broken. Major emphasis is placed on rehabilitation so that the employee can continue to make meaningful contributions to self, family, company, and society. Early detection of the problem by supervision is helpful, and extensive follow-up during the recovery stages is essential.

UNIONS FOR PROFESSIONALS?

We head this section with a question because the ultimate role of unions for professionals in industry is far from settled. We shall attempt to present both sides of the story as neutrally as possible, then close with our own opinions.

Union efforts to organize professionals have continued for many years, without much success. With the phenomenal industrial growth and prosperity of the 1950s and early 1960s, the climate was not conducive to union activity, and one of the earliest federations of engineering and scientific unions, the Engineers and Scientists of America (ESA), went out of existence in 1960.

In 1967, however, the AFL-CIO formed the Scientific, Professional, and Cultural Employees Council to stimulate union activity among professionals, their director stating that "a Ph.D. isn't worth an extra five T.V. stamps at the cashier's stand—and I think they are getting tired of seeing union laborers buy the steak while they buy the hamburger" (32). In 1968, former ESA members and other independent engineering unions formed the Council of Engineers and Scientists Organizations. Details of the organization, programs, and efforts of these councils are given in recent articles (33, 34). In 1973, the Oil, Chemical and Atomic Workers International Union formed a Professional Employees Division to serve the special needs of professionals in their industries. On request, this division distributes to professionals in relevant industries a free *Professional Advancement* newsletter (35). More stature was given to the activities of this division by its recent endorsement (35) by the Nobel laureates S. E. Luria, H. C. Urey, and Linus Pauling, who stated that the union organization of scientists and engineers should strengthen the economic security of those professions and enable them to act as responsible citizens as well as creative employees.

Why should professionals join unions and support union activities? What do unions offer that management will not provide on its own or that professional societies cannot offer? Among others, the things that professionals want include job security, higher salaries, recognition and respect as professionals, adequate benefits and retirement provisions, and challenging and meaningful job assignments that can be carried out under good working conditions. Advocates of unions cite the following reasons for professionals to affiliate with a chemical trade union (34, 36):

- Unions have demonstrated their ability to negotiate economic gains for professional employees.
- Unions are in a better position and have more experience than professional societies to encourage and lobby for the passage of legislation for improving job security, pensions, patent rights, and other benefits for professional employees.
- Comparison of the average incomes of physicians, dentists, and lawyers with those of engineers and scientists points out the need for collective bargaining through union representation. The relative bargaining strength of the individual is small.
- The interests of professionals are essentially different from the interests of their management.
- Professional employees are "trying desperately to preserve some of the attributes of the client relationship within an employer–employee situation" (36).

- Professional societies today are too management-oriented.

To support this point of view, an article in the *Professional Advancement* newsletter advocates (37) that professional societies and unions commit substantial sums to legislative action which would:

- "limit the right of any employer to discharge or harass an employee in retaliation for political action, ethically mandated conduct, or refusal to perform illegal work;
- "establish principles of corporate 'due process,' giving the employee the right to appeal any employment action he or she believes is unfair, arbitrary, or improper to a third party, after a fair hearing in the firm;
- "give workers the right to access, enforceable in courts of law, to any information relevant to their employment relationship, including for example access to one's own personnel files, access to personnel manuals and hiring, firing, and promotion guidelines. . . ."

It should be pointed out, however, that in some cases it is now unlawful for companies to provide special considerations and privileges to professional employees only.

One management consultant has attributed the need for professional unions directly to management action (38): "There would be no necessity for chemists and other professionals to join unions if managements were mindful of the fact that professionals have needs other than money. . . . The major reason for professional interest in unions is the basic feeling that their managements do not treat them decently, fairly, and honestly." For another statement of the same view see ref. 39.

Is this a representative feeling of chemists and engineers? How do the majority of the professionals feel? What are the counter-arguments?

A recent survey of the AIChE (40) indicated that most of chemical engineers interviewed felt that the union trend is not in keeping with the nature of the engineering profession, and that the individual engineer does not need a union to represent him. Only 4 out of 258 engineers interviewed stated that they would join a union, though more than a third did feel that unionization would probably be helpful for improving pensions, salaries, and overall job security.

An editorial in *Chemical & Engineering News* (41) comments that the needs for job security, benefit plans, and perhaps contracts "coincide with the needs of other workers, but for professionals, the answer is not union. Unions provide security in terms of maintaining status quo and providing organized clout counterbalancing owner-management which, in past days, criminally exploited labor. But there is good reason to believe unions are waning, their purpose largely dissipated as professional management has assumed many of their original functions." And an editorial in *Industrial Research* (42) is not against unions, but feels that the active technical society could protect the economic security of its members and should perform this function much better than a union.

Other arguments for the nonunion approach are that:

- The same benefits could be achieved by increased professional activities in the professional societies.
- Many of the benefits could be attained through cooperative efforts with industry, such as the Unified Guidelines.
- Unions may be satisfactory in specific situations, but do not offer a general solution.
- Professionals in a union would have to honor the picket lines of blue-collar workers.

Both the ACS and the AIChE officially maintain neutral positions regarding unions, neither supporting nor condemning their activities. It may become increasingly difficult for the professional societies to maintain neutrality in the future, as they and the unions compete to represent the best interests of professionals.

We close by stating our own feelings for what they are worth. Everyone recognizes that there are problems in employer–employee relations. To us, the most important fact is that everyone involved, including management, recognizes this, and steps are being taken to remedy many situations. While we are not opposed to unions in principle, we do feel that at this stage they offer very little to the professional employee that could not be obtained through increased cooperation between professional societies and company management. We see no reason to think unions could do a better job of representing the professionals. And we feel it would be a sad day when a prerequisite for professional employment would be union membership.

RELATIONS WITH HOURLY WORKER UNIONS

In the university, you may have done a fair amount of plumbing, electrical, and routine maintenance work yourself on your research equipment. In industry, except in very small companies, it is unlikely that you would be allowed to work on such a "do-it-yourself" basis, for a variety of good reasons. You certainly would not be if the company has hourly or wageroll worker unions. On joining such a company, you will find it important to become familiar with plant practices for getting things done. Equally important are the people involved—who to see when you need something, who are the bosses, the union representatives, and what is the seniority of the union people you will be working with. It goes without saying that a newcomer should treat everyone with respect, be a good listener, and try to understand the motivations and concerns of the union workers with whom he is in contact.

As a beginner, you will probably not be involved in direct dealings with union representatives in grievances or contract negotiations. There are guidelines for union relations to which you may wish to refer later (43). If there is a strike, with picket lines, at your plant, your supervision will advise you on what to do, which may include filling in outside your regular duties in order to keep the plant going.

HIRING AND LAYOFF PRACTICES

These topics have received much attention recently, especially as part of the guidelines presented in Appendix I and the corresponding ACS guidelines (23), and the activity of the ACS Committee on Professional Relations. The ACS recently published summaries of the termination conditions of employers heavily involved in layoffs of chemists and engineers (44), and has indicated that it will continue to publish this sort of information. (Recall that the ACS has chosen not to endorse the intersociety guidelines, the main points of disagreement being in these termination conditions where the ACS guidelines are more specific.) In the published summaries, company practices were compared to the ACS guidelines in these categories:

Advance notice (the ACS recommends a minimum of four weeks); severance pay (two weeks' salary for each year of service); assistance (every effort, including retraining, to find another position within the company); pension vesting (full after 10 years); involuntary retirement (employees retired in this manner should be treated at least as well as those laid off for economic reasons); employee service (no one with more than 10 years service should be terminated unless there is documented evidence of inadequate preformance, or for cause); employee protection plans (should continue for at least one month after termination); and rehiring privileges (explain fully to the employee, and rehire those laid off for economic reasons before recruiting new employees).

Fortunately, as the published comparisons showed, most companies currently do a good job in meeting these guidelines.

THEFT

Like shoplifting in retail stores, the pilfering of small items by employees is a major problem to industrial concerns. In addition to the usual security precautions, companies have been forced to take legal action against some employees to protect their—and their stockholders'—interests.

It is difficult for us to see why the practice of taking office supplies, small tools, and the like has proliferated. It is *not* a "fringe benefit"; it is out-and-out stealing; and it's *not* true that "they don't worry about such small things"; they must, or the problem will mushroom. As a professional, you should be above such petty thievery and not condone it in others.

PENSIONS AND RETIREMENT

As we said in Chap. 3, it used to be that a good way to "turn off" a campus interviewer was to inquire about pension and retirement policies. But times have changed, and it is quite likely that you did consider these things in selecting your present company. Among things you will have going for you at retirement, whether at age 65 or earlier, will be your pension, company life and health insurance, and Social Security. Your company literature will give you details on these. We urge you to supplement them by starting now to build your own sound financial program of savings, stocks, insurance, home, and all the other things that will contribute to the good life at the end of your working career.

SHARPENING YOUR SKILLS

As a professional, you will find it necessary to work hard and continuously to avoid technical obsolescence. It is essential that you maintain your scientific and engineering competence, your creative productivity, and your ability to communicate your ideas and discoveries to others such as supervisors, associates, clients, and the public. In this section we explore three areas pertinent to improving your skills, and attempt to answer such questions as:

- Why should you keep up-to-date?
- How can you keep pace with the rapid advancement of science and technology?
- How can you keep informed about the latest developments in your profession and your industry?
- How can you improve your ability to communicate?

PROFESSIONAL SOCIETIES AND TRADE ASSOCIATIONS

A professional society is an organization of individual members of a specific profession devoted to common goals, such as upholding the profession and serving the technical, moral, ethical, and economic interests of its members for the benefit of society. A trade association, on the other hand, is composed of member companies—competitors within the same industry—who join together to formulate and discuss the policies of that industry in areas of mutual concern, within the framework of such legal constraints as the antitrust laws. We have listed the names and addresses of a few professional societies and trade associations in Appendix IV, and encourage you to explore those you find particularly interesting. There are vast numbers of these—almost 20,000 trade associations alone (45)—so you will understand if we have missed listing your favorite.

The three largest professional societies for chemists and chemical engineers are those we have mentioned many times already, the American Chemical Society (ACS), the American Institute of Chemical Engineers (AIChE), and the American Institute of Chemists (AIC). Their combined membership in 1974 was around 160,000. Their news journals, sent to all members, are (respectively) *Chemical & Engineering News, Chemical Engineering Progress,* and *The Chemist.* Like all professional societies, they will be happy to provide free information on membership requirements and activities.

We have already described some of the activities of the professional societies, such as aiding professionals seeking employment, developing codes of ethics, and establishing employment guidelines. They also lead the way in disseminating technical information through books, journals, and other media, and in reporting on the current status of the professions they serve. We highly recommend that you join at least one professional society, take part in supporting its programs, and attend its technical meetings. Remember too that since you are a professional, your dues are tax deductible.

The nearly 20,000 trade associations in the United States employ over 30,000 executives and staff specialists. About 30 national trade associations are active in the chemical industry; reviews of their functions are available (45). The oldest and most diversified of these is the Manufacturing Chemists' Association (MCA). Founded in 1872, at its centenary its 173 member companies represented over 90% of the production capacity of basic industrial chemicals in the United States and Canada.

The MCA (46) is composed of four operating departments: technical, government relations, public relations, and information services. Areas of its involvement range from analyzing broad economic policy issues and problems affecting the chemical industry to operating an around-the-clock "hot line" for emergencies involving chemicals in transportation accidents. You may already be familiar with the "Chemical Safety Data Sheets" and the "Safety Guides" developed and widely distributed by the MCA. We suggest that you become familiar with the trade associations serving your industry and, where appropriate, help your company in its support of their activities.

CONTINUING EDUCATION

A convenient parameter for describing technical obsolescence is the half-life of professional competence, the time it takes for a practicing professional to lose half his competence. Considering that your industrial career can be expected to last 35–40 years, it may come as somewhat of a shock to learn that this half-life for technical competence was estimated (47) to be only *five years* for the 1971 engineering graduate. This is down drastically from 12 years in 1940, due in part to the rapid pace of scientific and technological advances.

Bearing this situation in mind, we hope you will pardon us for asking even the newest graduates whether you are already feeling the danger signs of obsolescence (48). Some of the symptoms are:

- It is increasingly difficult to read and understand technical papers in your field.
- You have less understanding of, and more difficulty in applying, new problem-solving mathematical techniques.
- New assignments and tasks seem increasingly difficult.
- New technical concepts appear confusing.
- Contemporaries seldom seek your advice any more.
- Family or outside activities seem to take all your time away from the job.

Even if these danger signs don't apply to you now, it's time to take action to keep them away. *You* must do this—no one can do it for you—and it takes continuing effort, best put into high gear as soon as your formal education is behind you. Don't get caught in the trap of having the best intentions but never following through, and remember that professional competence requires much more than just acquiring new knowledge. That knowledge is useless if it can't be properly applied, and the results properly communicated to others.

Here are some things you should be doing to avoid technical obsolescence. Companies generally support and encourage these activities, but they usually leave to you the choice of what suits your needs best:

- Join your professional society, and through its news journal keep up to date on its activities and the changing situations in science and technology in industry, society and government.

- Read the major technical journals in your field regularly, following a fixed schedule to ensure that you don't lag.
- Visit your technical library regularly to become familiar with new books in your field, review recent government publications, chemical abstracts (including patents) and other specialized abstracts, and scan technical journals other than those you normally read.

In connection with the preceding two points, note that about 20% of a professional's working time should be devoted to reading in order to keep abreast of new publications. You may wish to reduce this time by taking a speed-reading course.

- Participate in continuing-education programs offered by the professional societies and many universities (49). Among the former are the ACS Short Courses, Audio Courses, and Famous Scientists tape cassettes, the AIChE Management Seminars and Today Series, and the American Association for the Advancement of Science Audiotapes programs.
- Attend local and national technical meetings of the professional societies regularly. Most companies allow two trips a year to attend such meetings.
- Take advantage of evening courses, seminars, and the library at your local college or university. Most companies pay the tuition costs for courses you take.
- Attend trade shows periodically.
- Participate in on-the-job opportunities for continuing education, such as special courses (49), programs in cooperation with local colleges, and seminars by consultants and visitors.

In the recent book, *Chemistry in the Economy* (50), the ACS emphasized the importance of increased effort on the part of industry, universities and colleges, and scientific and professional societies to meet the needs of continuing education for today's scientists and engineers. Specifically, it recommended that companies "make even greater use of postdoctoral arrangements, university extension classes, selective tuition assistance, sabbaticals, special and regular seminars, visiting speakers, in-house courses, closed-circuit television, book purchases, library services, and other means of maintaining, upgrading, and extending the skills" of professionals.

COMMUNICATION

By now you must be aware that we make a very strong point of the need for the professional to master the skills of communication. We've seen too many engineers or scientists who can't put two sentences down on paper without such atrocities as bad grammar, misspelled words, or clichés, for us to pass over the importance of this need. And we aren't alone. The ACS recommends in *Chemistry in the Economy* that university chemistry programs should contain strong academic requirements for written and oral communication, and even more important—that practice and experience in oral reporting and in writing technical reports should be made a part of the chemistry curriculum itself. The complaints are echoed in other disciplines, too.

Are we talking to *you*? Do *you* need to improve your communication skills? Try asking yourself these questions:

- Can I get my ideas across to others easily and accurately?
- Do I find it hard to talk to the boss?
- Does my supervisor have to correct my reports and letters or ask me to rewrite parts of them?
- Can I organize my thoughts effectively, pulling the important points out of the details with ease?
- Do I "choke up" when required to speak before an audience?
- Am I afraid to ask a question of the speaker in a big meeting?

We'll wager that if you are honest you will see places where improvement could be made. Do something about it! Remember that your ability to communicate effectively can be an important factor in getting ahead on the job.

We can't teach you how to communicate in a few paragraphs, but we can emphasize the importance of practice and experience in this as well as many other areas. Here we offer some specific suggestions:

Oral communication in which you must participate includes talking to your boss and your colleagues, reporting at on-the-job meetings, speaking as a company representative at technical meetings or in public, and using the telephone properly. If you have difficulty in these areas, try to work with someone else who can offer constructive criticism and help you define and overcome your specific problems. Set up realistic goals for self-improvement and work toward them at a reasonable rate. Consult the literature (51–56), especially at the start, to make sure you are taking the right approaches. Possibly you may wish to join your local Toastmaster's Club, where you can gain experience and learn how to command your audience's attention.

Sometimes we think that the telephone is an invention of the devil, and wish we could tear it out by the roots. Don't misuse, or overuse it, even though it offers a good way of getting many things done. Especially for long-distance calls, where time is money, organize your conversations in advance to be as concise as possible, and follow company guidelines.

In the realm of written communications, you will need to prepare letters, summaries for management, technical and business reports, patent and research proposals, papers for publication, and even to edit the writing of others. You can save valuable time by learning how to combine your oral and written communication skills through the use of dictation.

In addition to learning and adhering to the accepted rules of spelling and grammar, you can greatly improve your technical writing by attention to that ephemeral quality of style (57) and to careful organization. For the latter, making an outline is essential before you set about the actual writing. For very short pieces, this may be no more than a mental exercise, but if any amount of detail is involved, get the outline down on paper. As you do so, define the purpose and objectives of the writing, and organize the outline to meet them in the clearest and most direct way. Your company will probably have guidelines on the content and arrangement of internal technical reports, and papers prepared for publication should be written to follow the style of the journal for which they are prepared. The literature (58–62) has many useful examples, and short courses on technical writing offered by universities or professional societies are highly useful.

Why is there so much difficulty in learning how to communicate skillfully? We suppose it goes back to bad habits begun in grammar school, and we certainly don't hold any illusions about reforming the educational system. But there *was* a reason for all the drill and exercise, the composition writing you were supposed to do in high school. We are continually saddened and dismayed to see how little effect it had on the students we get in college, at the graduate as well as undergraduate levels. But the problem is real, and we urge you to meet it head on.

SHARED RESPONSIBILITIES

Safety, the preservation of the environment, and keeping the best interests of society in mind are everybody's responsibilities, no less in industry than anywhere else. Each company, group, and individual must do his part in order to ensure the good of everyone. The concern for safety at all these levels has recently been given great impetus by the passage of a group of federal regulations designed to ensure safe and healthful working conditions for all employees.

The wanton exploitation of our natural resources has led us into a worldwide energy crisis, and the pollution of our environment has led to the passage of regulatory legislation, the creation of such government arms as the Environmental Protection Agency, and a significant increase in the activity of consulting firms dealing with environmental engineering. All these evidence the commitments now being made to the solution of environmental problems, and there is no doubt that the chemical industry will play a major role in these efforts.

SAFETY AND HEALTH

It might be expected that because of the hazardous nature of many chemical operations, the risk of injury in the chemical industry would be unusually high. This is not the case. In fact, the injury frequency rate (lost-time injuries per million man-hours worked) was far lower in the chemical and allied products industries (8.5) than in industry as a whole (15.2) in 1970 (63). And the Du Pont Company, always a leader in its emphasis on safety, recorded the lowest frequency rate (0.16) in its history in 1972.

These good records don't just happen; they are achieved through continuing hard work by everyone involved with the chemical industry. Companies, professional societies, trade associations, unions, and individual employees all contribute. Safety is never taken for granted.

Most large companies have active safety departments and industrial hygiene groups which coordinate these activities on a companywide basis. In companies paying adequate attention to safety (and showing it in their records), safety meetings, programs, and inspections are regularly scheduled and involve all employees, not just those in research, development, fabrication, and manufacturing areas. The emphasis on safe and healthful operations is important to all, and all are required to do their part. Furthermore, safety doesn't end at quitting time, and the carry-over of on-the-job job safety to home and vacation work and play is an important part of ensuring the maximum in performance and well-being of every employee. You will be expected to comply with all safety and health regulations of your employer, and as a professional, to provide a good example at all times. One of these regulations may be to report promptly to your safety group all accidents, even minor ones and "near misses" causing no lost time. It's important to comply, for your own sake as well as that of your fellow workers.

THE OCCUPATIONAL SAFETY AND HEALTH ACT (OSHA)

The Williams-Steiger Act, commonly known as OSHA, was passed in 1971 "to insure so far as possible every working man and woman in the nation safe and healthful working conditions and to preserve our human resources" (64, 65). The act provides for an Occupational Safety and Health Administration to establish national standards for all industries, to conduct inspections to ensure that these standards are met, to levy fines and penalties for violations, to cooperate with other agencies at all levels, and to offer assistance in compliance by ensuring that its policies are correctly interpreted and implemented. In 1971, OSHA inspectors made almost 33,000 visits to just under 30,000 plants employing 5.9 million people. Some violations of the law were noted in more than 75% of these plants, and over 100,000 citations were issued (66).

OSHA gives the employee (or, if there is a union, his representative) several important rights (67). If he believes there is a violation of a job safety or health standard or an imminent danger, he may request an inspection (anonymously if desired and without danger of possible discharge or discrimination). His representative, if any, may accompany the inspector during the inspection. If the employer protests a citation, the employee can participate in the resulting hearing. He may observe the company's monitoring procedure for toxic or harmful agents, and his representative may request an investigation of toxicity of any substance found in the working place.

The first major group of toxic substances to be investigated by OSHA was carcinogens (68). Under pressure from the Oil, Chemical, and Atomic Workers Union and the Health Research Group (a Ralph Nader affiliate), OSHA quickly established standards for safe exposure levels for 14 of these chemicals, including some dyestuff intermediates and antioxidants. The list is expected to grow substantially: NIOSH, the National Institute for Occupational Safety and Health (the research arm of OSHA), now lists more than 600 carcinogens, and a recent article (69) claims that "chemicals—in the work place, in the environment, and in the diet—may be the single most important cause of human cancers."

THE ENVIRONMENT

The 1970s may well go down in history as the era of pollution abatement in the United States. Legislation has provided for the establishment of environmental standards for all forms of pollution, and federal, state, and local agencies, from the Environmental Protection Agency on down, have been established to implement them. The task of cleaning up our environment and reducing industrial and municipal pollution to acceptable levels is herculean. The problems are very complex, and their solution will require not only scientific and engineering effort, but also the delicate balancing of the needs of today's society with the overall costs and long-range benefits involved. Solutions won't be found overnight, but they must be found if our way of life is to be preserved.

They're everybody's problems, and they can only be solved with plenty of dollars, technology, and talent.

The estimated cost of the job is staggering, and increasing rapidly (70). The chemical and allied products industry spent only $0.13 billion for pollution control in 1971, but is expected to invest $1.3 billion in capital outlay alone between 1972 and 1976. In the paper industry it's the same story—$0.3 billion in 1972, $3.3 billion capital outlay predicted for 1972–76. And there are many more industries plus all the municipalities yet to be counted.

In most cases, the technology needed to stop the pollution exists. The problem is to balance its application with respect to costs and to the present and future needs of all. Policy makers at all levels must be educated to ensure the right decisions, and the public must be educated to the nature of the problems and the steps being taken to solve them. Without education and sensitivity, decisions may be based on emotion rather than wisdom—something we cannot afford.

This is where you, the professional, come in. You and your colleagues must supply the talent that will ensure that the job gets done right. You must see that you are educated and fully up to date on those aspects of pollution abatement that affect you, both on and off the job. Then you must apply your knowledge to educate your family, community, employer, and legislators.

The education that you obtain and pass on must combine ethical judgments with technological decisions, and must be consistent with legal obligations. It is difficult to separate the facts from the editorializing, but some recent booklets (71) are helpful. Become familiar with codes and standards that affect you on your job, and with what is being done in your community to reduce pollution. Your company and public libraries will have plenty of books on the subject these days. Pay particular attention to problems with ethical connotations, though they are very knotty and so varied as to be difficult to define. Working from within your company to investigate these is usually the proper approach. A prize-winning series of recent articles (72) is appropriate.

Apply your technical background in dealing with environmental problems outside of your job whenever possible. Communicate your concerns to your family and your community. Inform them of what is being done, and urge both your company and your professional society to continue to educate the public on programs for pollution abatement.

Because chemicals pollute, the general public tends to equate pollution with the chemical industry. In fact, the transportation and energy industries are by far the largest contributors, at least to air pollution. To inform the public of the true position of the chemical industry and its achievements in pollution abatement will require continuing education of a very positive nature. The record is clear; let's help to get it across.

THE PUBLIC

In the last few years companies, professional societies, and individuals have become increasingly aware that they owe a responsibility to the public. A number of well-known professionals have expressed this in the following statements:

- "The cry for social responsibility demands that scientists consider society their client and the public interest their paramount concern . . ." (73).
- "In a word, we should work toward a more perceptive and participatory social conscience for chemistry and chemists" (74).
- "We must change from [an] attitude of exploitive chemistry to a new humanistic chemistry" (75).
- "Scientific and educational societies should take an active interest in public and political affairs . . . not only because the work of their members is affected . . . [but] because the influence of that work on society is affected" (76).
- "The apparent direction of evolution of our society suggests that public position may become almost as important to an industrial corporation . . . as economic condition" (77).

Clearly, the call has been made for recognition of responsibility to the public and action to implement it. A recent series of articles (78) supplies some of the answers and includes specific suggestions for the professional's action in response to it. Among these courses of responsible action are running for public office, working for local civic groups, writing to legislators, becoming active in a professional society's environmental or other public-service group, providing tutoring or career guidance to students (especially the underprivileged), serving on school or hospital boards, and many, many more. In this wide spectrum of opportunities there should be some to interest and challenge any self-respecting engineer or scientist. By accepting this public responsibility he—you—can become a better citizen as well as a better professional.

REFERENCES

1. Anon., "Professional Standards," American Institute of Chemical Engineers, New York (no date).

2. J. Roger O'Meara, "Employee Patent and Secrecy Agreements," Studies in Personnel Policy No. 199, National Industrial Conference Board, New York, 1965.
3. Anon., "Employment Agreements," Professional Handbook Series, American Chemical Society, Washington, D.C. 1971.
4. Anon., "Guidelines to Confidentiality Clauses," American Institute of Chemical Engineers, New York (no date).
5. William J. Bailey, "Employment Contracts," *Chemtech* **4**, 707 (1974).
6. J. R. O'Meara, "How Smaller Companies Protect Their Trade Secrets," Report No. 530, The Conference Board, New York, 1971.
7. T. J. Walsh and R. J. Healy, *Protecting Your Business Against Espionage,* American Management Association, New York, 1973.
8. Anon., "Trade Secrets—The Technical Man's Dilemma," *Chem. & Eng. News* **43** (3), 80 (Jan. 18, 1965).
9. W. Wade, *Industrial Espionage and Mis-Use of Trade Secrets,* Advance House, Ardmore, Pa., 1964.
10. Thomas H. Arnold, Jr., "Are You Locked into Your Job by What You Know?" *Chem. Eng.* **73** (10), 141 (May 9, 1966).
11. R. Ellis, *Trade Secrets,* Baker, Voorhis, New York, 1953.
12. Anon., *Trade Secrets . . . Ethics and Law,* American Chemical Society, Washington, DC., 1968.
13. G. Frederic Holden, "Ethics in Industry—The Dilemma," *Chem. Eng.* **80** (28), 130 (Dec. 10, 1973).
14. Anon., "Professional Ethics for Chemists," Committee on Professional Relationships, American Chemical Society, Washington, D.C., 1972.
15. Anon., "The Chemist's Creed," American Chemical Society, Washington, D.C., 1965.
16. Anon., "The Canon of Ethics of the American Institute of Chemical Engineers," American Institute of Chemical Engineers, New York (no date).
17. Anon., "The Code of Ethics of the American Institute of Chemists," *The Chemist* **15** (1), 19 (1938).
18. G. H. Marcus, W. H. McCarty, T. B. Richey, and H. S. Rossman, "Some Opinions on Employer-Employee Relations," *Chem Eng. Prog.* **69** (2), 37 (1973).
19. Anon., "Issues of Professionalism and Employment," *Chem. Eng. Prog.* **69** (11), 37 (1973).
20. L. V. McIntire and M. McIntire, *Scientists and Engineers—The Professionals Who Are Not,* Arcolo Communications, Lafayette, La., 1972.
21. Anon., "Unified Guidelines Endorsed by AIChE," *Chem. Eng. Prog.* **69** (3), 15 (1973).
22. Anon., "Guidelines to Professional Employment for Engineers and Scientists," 1973 (available from endorsing professional societies, including AIChE).
23. Anon., "ACS Guidelines for Employers," *Chem. & Eng. News* **50** (21), 25 (May 22, 1972). In April, 1975, the ACS approved new, revised Professional Employment Guidelines, and made them available in booklet form.
24. R. P. Kostka, "Why Chemical Engineers Want Guidelines," *Chem. Eng. Prog.* **68** (11), 22 (1972); C. E. Johnson, "A Company's Viewpoint," ibid., p. 26; H. Popper,

"Alternatives to Guidelines," ibid., p. 29; Anon., "What the Discussion Brought Out," ibid., p. 32.

25. Anon., "Attitudes Vary Widely on ACS Guidelines for Employers," *Chem. & Eng. News* **50** (46), 38 (Nov. 13, 1972).
26. B. F. Somerville, "Employment Guidelines Widely Endorsed," *Chem. & Eng. News* **51** (20), 6 (May 14, 1973).
27. J. R. Catenacci and P. H. McNamara, "Grass-Roots Guidelines Activity," *Chem. Eng. Prog.* **70** (2), 28 (1974).
28. P. J. Davey and J. K. Brown, "The Corporate Reaction to 'Moonlighting'," *Conf. Board Rec.* **7** (6), 31 (1970).
29. H. M. F. Rush and J. K. Brown, "The Drug Problem in Business," *Conf. Board Rec.* **8** (3), 6 (1971).
30. S. Habbe, "The Drinking Employee—Management's Problem?" *Conf. Board Rec.* **6** (2), 27 (1969).
31. Anon., "Alcoholism: New Victims, New Treatments," *Time* **103** (16), 75 (April 22, 1974).
32. Anon., "AFL-CIO Has Formed a Council to Stimulate Union Activity Among Professional Employees," *Chem. & Eng. News* **45** (14), 29 (March 27, 1967).
33. Anon., "Move to Organize Professionals Accelerates," *Chem. & Eng. News* **46** (52), 24 (Dec. 9, 1968).
34. Joan M. Nilsen, "Big Union Target—You," *Chem. Eng.* **80** (29), 24 (Dec. 24, 1973).
35. *Professional Advancement,* Professional Employees Division, Oil, Chemical, and Atomic Workers Union, Denver, Colorado **1** (1), (Nov. 1973).
36. J. Golodner, "Unions for Professionals," *Chemtech* **2**, 133 (1972).
37. Peter Petkas, "How to Give Meaning to Professional Responsibility," *Prof. Adv.* **1** (2), 2 (February 1974).
38. A. A. Imberman, "What Kind of Executives Cause Labor Trouble?" *Chemtech* **2**, 272 (1972).
39. Anon., "Management Error: A Cause to Unionize," *Chem. & Eng. News* **50** (22), 2 (May 29, 1972).
40. Anon., "Attitudes Towards Unions," *Chem. Eng. Prog.* **70** (2), 41 (1974).
41. P. P. McCurdy, "Union Is Not the Answer," *Chem. & Eng. News* **50** (19), 3 (May 8, 1972).
42. A. F. Plant, "The Third Call," *Ind. Res.* **15** (10), 7 (October 1973).
43. George A. Kauber, "Hourly-Workers Unions and the Engineer," *Chem. Eng.* **81** (1), 118 (Jan. 7, 1974).
44. A. C. Zettlemoyer, "Committee Examines Company Layoff Practices," *Chem. & Eng. News* **51** (32), 26 (Aug. 6, 1973); anon., "More Company Layoff Data Disclosed," *ibid.* **52** (11), 18 (March 18, 1974).
45. D. M. Kiefer, "The Trade Association: The Chemical Industry's Multitongued Voice," *Chem. & Eng. News* **45** (6), 115, (Feb. 6, 1967).
46. Anon., "MCA—What It Is—What It Does," Manufacturing Chemists' Association, Washington, D.C., 1972.
47. S. S. Dubin, "The Psychology of Keeping Up-to-Date," *Chemtech* **2**, 393 (1972).
48. H. G. Kaufman, *Obsolescence and Professional Career Development,* American Management Association, New York, 1974.

49. Howard J. Sanders, "Continuing Education," *Chem. & Eng. News* **52** (19), 18 (May 13, 1974); (20), 26 (May 20, 1974).

50. *Chemistry in the Economy,* American Chemical Society, Washington, D.C., 1973.

51. E. M. Kipp, "Communicating Better in Research and Engineering," *Chem. Eng.* **79** (17), 75 (Aug. 7, 1972).

52. W. Donald Lieder, "Challenges and Rewards of Public Speaking," *Chem. Eng.* **80** (3), 88 (Feb. 5, 1973).

53. Dennis J. Chase, "Conferencemanship," *Chem. Eng.* **80** (12), 118 (May 28, 1973).

54. James Grayson Ford, "Oratory Isn't Dead But Many Speakers Are," *Chem. Eng.* **80** (11), 150 (May 14, 1973).

55. Hoyt S. Gramling, "Going to a Meeting? Why?" *Chem. Eng.* **80** (1), 130 (Jan. 8, 1973).

56. M. Ivens, *The Practice of Industrial Communication,* Business Publications, London, 1963.

57. W. Strunk, Jr. and E. B. White, *The Elements of Style,* Macmillan, New York, 1959.

58. H. M. Quackenbos, "Creative Report Writing—Part I," *Chem. Eng.* **79** (15), 94 (July 10, 1972); "Creative Report Writing—Part II," *ibid.,* (16), 146 (July 24, 1972).

59. H. Hoover, *Essentials for the Technical Writer,* Wiley, New York, 1970.

60. S. Jordon, J. M. Klineman, and H. L. Shimberg, eds., *Handbook of Technical Writing Practices,* Wiley-Interscience, New York, 1971.

61. H. J. Tichy, *Effective Writing for Engineers • Managers • Scientists,* Wiley, New York, 1966.

62. S. Mandel, *Writing for Science and Technology,* Dell, New York, 1970.

63. Anon., "Job Safety and Health Program Takes Shape," *Chem. & Eng. News* **51** (25), 17 (June 18, 1973).

64. *The Williams-Steiger Act,* Federal Register, May 29, 1971.

65. Anon., "All About OSHA," OSHA Publication 2056, U.S. Department of Labor, Washington, D.C. (no date).

66. Anon., "Occupational Safety and Health Act," *Chem. Eng., Databook Issue* **80** (5), 13 (Feb. 26, 1973).

67. F. K. Foulkes, "Learning to Live with OSHA," *Harvard Bus. Rev.* **51** (6), 57 (Nov.–Dec. 1973).

68. Anon., "Final Rules Set for Exposure to Carcinogens," *Chem. & Eng. News* **52** (6), 12 (Feb. 11, 1974).

69. T. H. Maugh, II, "Chemical Carcinogenesis: A Long-Neglected Field Blossoms," *Science* **183**, 940 (1974).

70. Joan Nilsen, "Cleanup: What's It Worth?" *Chem. Eng.* **79** (13), 48 (June 12, 1972).

71. Anon., "71 Things You Can Do to Stop Pollution"; anon., "What Industry Is Doing to Stop Pollution," Keep America Beautiful, Inc., New York (no date).

72. Herbert Popper and Roy V. Hughson, "How Would *You* Apply Engineering Ethics to Environmental Problems?" *Chem. Eng.* **77** (24), 88 (Nov. 2, 1970); Roy V. Hughson and Herbert Popper, "Engineering Ethics and the Environment: The

Vote Is In!" *ibid.,* **78** (5), 106 (Feb. 22, 1971); "Environmental-Ethics Panel Offers Views and Guidelines," *ibid.,* **78** (6), 109 (March 8, 1971).

73. Michael Jacobson, "Science in the Public Interest," *Chem. & Eng. News* **50** (3), 1 (Jan. 17, 1972).
74. F. A. Long, "Social Conscience of Chemists," *Chem. & Eng. News* **50** (14), 3 (April 3, 1972).
75. Robert H. Linnell, "Humanistic Chemistry," *Chem. & Eng. News* **49** (47), 3 (Nov. 15, 1971).
76. Richard L. Kenyon, "Scientific Societies and Public Affairs," *Chem. & Eng. News* **45** (4), 5 (Jan. 23, 1967).
77. Richard L. Kenyon, "Industry and the Pecking Order," *Chem. & Eng. News* **45** (8), 5 (Feb. 20, 1967).
78. Herbert Popper, "The Chemical Engineer: Society's Problem-Maker or Problem-Solver?" *Chem. Eng.* **79** (13), 78 (June 12, 1972); "Running for Public Office: How? Why? Why Not?" *ibid.,* (15), 84 (July 10, 1972); "Winning Friends and Influencing People," *ibid.,* (18), 109 (Aug. 21, 1972); "Creating New Educational and Job Opportunities," *ibid.,* (21), 159 (Sept. 18, 1972); "Questions, Answers, and Conclusions on the Ch.E. and Society," *ibid.,* (24), 111 (Oct. 30, 1972).

CHAPTER 5

ADVANCEMENT

What does "advancement" mean to the young industrial employee? It can mean any of a number of things, depending on personal goals and temperament. Some people are very ambitious and thrive in "pressure situations" that would lead others to an ulcer or a nervous breakdown: These hard drivers are likely to be anxious to get into supervision and direct the work of others. Some wish to acquire more responsibility and autonomy in technical and scientific matters without inheriting the burdens of supervision: For these, advancement implies climbing the scientific ladder. Others may be content with the satisfaction of doing their job well without accepting more responsibility: To them, the recognition and respect of supervisors and associates may be adequate reward.

Regardless of the differences in goals, recognition and respect for the job being done are key ingredients affecting performance and satisfaction on the job, regardless of level. Salary increases are, of course, a significant and tangible part of job recognition.

Perhaps, in your job interview, you were asked about your long-range goals: What do you see yourself doing ten years from now? This is a difficult but very pertinent question, one that needs to be repeated periodically. Most people—not only those in industry—find that the answer changes frequently over the course of their career. Finding the answer in your job is not easy, particularly at the beginning. It takes time to get the picture, to recognize the pattern of progress and advancement within a company, and to adjust the goals of your career to it. Often, progress is influenced by many small factors, some of which cannot be controlled by the individual. Despite the level of importance of advancement opportunities in your employment decision, the need to reassess them always remains. It is important, therefore, to get into the habit of establishing and reassessing meaningful and realistic goals for career advancement.

There is no single formula for achieving such goals, no matter how realistic, unless it is one of continuing hard work. You may find it both amusing and interesting to read or reread some of the best-selling books (1–3) on routes to corporate success, including two (4, 5) recently published by the American Management Association.

PERFORMANCE APPRAISAL AND CAREER DEVELOPMENT PROGRAMS

One of the first things that you will find out as your industrial career gets under way is that your progress and advancement are being carefully monitored by your supervisor. Periodic reviews between you and your boss are used by most companies to provide a formal opportunity for the discussion of these subjects. We start our discussion of advancement with this regular checkpoint.

THE PERFORMANCE REVIEW

At regular intervals—often twice a year for the first few years and once a year thereafter—you will be asked to review your performance for the past period with your supervisor. Among the objectives of this performance review are:

- To serve as a basis for your promotions, raises, transfers, new assignments, etc.
- To help you recognize how you fit into the organization, assess your chances for advancement, and delineate the areas in which self-improvement is desirable.
- To provide a basis for communication between you and your supervisor, leading to a better understanding of mutual concerns and the establishment of a meaningful career development program.
- To provide a consistent basis for management review and comparison of the performance of all employees.

The major results of the performance review are a merit rating and a career development program. The merit rating tells you where you stand, relative to the standards expected by the company. The career development program is designed to help you set and attain realistic goals.

Usually your supervisor will initiate the performance review process by filling out a performance evaluation form. He may do this by himself, or call on other supervisors who know you for assistance. His analysis is frequently reviewed by his boss or others before it is put in final form.

Performance evaluation forms vary widely in length and complexity. The following description, taken in part from ref. 6, is typical. The form is in two major parts: a performance appraisal section filled out in advance by the supervisor, and a career development portion filled out during or after the interview with the employee. In the performance appraisal section, you are rated in a number of categories on a scale ranging from poor to superior in, say, five steps. Among the rating categories may be quality of work (accuracy, completeness, efficiency), quantity (output of satisfactory work), job knowledge, adaptability (ability to accommodate to changed conditions), creativity (originality), attitude (open-mindedness and cooperation), communication (oral and written skills), leadership, safety and housekeeping, attendance (health and absenteeism), dependability, and, finally, an overall rating.

To assist your supervisor in rating you in each of these categories—and you can see how their subjective nature can lead to many problems—a descriptive phrase often accompanies each combination of category and rating. For example, an average rating of job quality might be described as "consistently meets job standards"; a superior rating for job knowledge as "consistently shows superior understanding of all phases."

This portion of the appraisal form may conclude with a series of questions for your supervisor to answer, such as:

- Describe any other factors affecting the employee's performance—for example, tact, poise, common sense, personal and professional integrity.
- In what specific areas do you feel improvement should be made?
- What actions do you plan to suggest to help bring about these improvements?
- Describe any change or improvement in performance since the last appraisal.

Filling out the performance appraisal form is a very tough job, and supervisors find it one of the heaviest responsibilities their position requires. Some of the problems that aren't quite so obvious are how to maintain consistency throughout a large company, with many employees in various groups; and what to do about the really good employee who consistently rates near the top in each category of his job description but continues to improve every review period.

When this portion of the form has been completed, your supervisor will schedule an interview with you to go over it and to formulate programs to help you in any areas where this is needed. This interview should be approached objectively, with an open-minded and receptive attitude, for when accepted in this way the merit rating process can be a valuable tool in your program for achieving your goals.

As the final step in the performance review, the supervisor records the results of your interview for his report to management and the permanent records. Often this is done on still another form by answering questions such as these:

- In what areas was it agreed that the employee should try to improve?
- What career development activities were planned with the employee to help him achieve this improvement?
- What did the employee say about his goals, desires, job preferences, self-improvement activities, etc.?
- What are the employee's greatest strengths (abilities, personal qualifications, relationships with others, activities in which he excels)?
- In what respects can he further prepare for future assignment (qualities to be strengthened, relationships with others, improvement of capabilities)?
- What specific action is planned or recommended to help him prepare for future assignment?
- For what type of work does he express personal preference?
- List other information of value in placement consideration (outside activities, courses completed, special talents, etc.)
- Describe any health or personal considerations of employee or his family that might affect his future career.

At the end of this section there is usually space for your supervisor to indicate his recommendation for timing and level of your next promotion, and to add any comments he may have to amplify the foregoing statements.

Although we believe the description we have given of the performance review is typical, there are many variations, and you may wish to consult some books on the subject (7–10) if it interests you.

THE SUPERVISOR'S ROLE

Your supervisor's role in furthering your career is not limited to his part in the formal performance review. Much depends on the nature of your relationship with him, at all times. With several possible types of supervisors (9)—the feared but respected manager, the nice (but slightly incompetent) guy, the "over-the-hill" supervisor, or the incompetent one, as well as the one with whom an easy give-and-take relationship can be established—it may require all your skill to find and maintain the proper attitude. It may be helpful for you to rate your supervisor (Fig. 15), and yourself as well, using the form given in ref. 11.

You will find life on the job much easier if you recognize that perform-

"Why, thank you, sir, and I had it in mind to tell you what a bang-up job I think you're doing."

FIGURE 15. From *The New Yorker* **50**(47), 31 (January 13, 1975). Drawing by Mulligan. Copyright © 1975 by The New Yorker Magazine, Inc.

ance improvement suggestions are made for your benefit, and assume the proper attitude toward them. Some factors which may affect your receptivity to these suggestions are listed in Table 12. An attempt to adopt positive attitudes would be well worth while.

There is good material in the literature (12, 13) on what the supervisor as well as the employee can do to improve the relations between them. The suggestions given in these articles are good for self-rejuvenation when it is needed, too.

GOAL SETTING

Many companies place considerable emphasis on the establishment of attainable goals at all levels of the organization. These goals are often used for budget planning and establishing management commitments, as well as for judging both individual and company performance. In such a company you will be required to establish goals, and how closely

TABLE 12

Some Common Factors Affecting Receptivity to Performance Improvement Suggestions*

The employee is *more* likely to be receptive to his manager's performance improvement suggestions *if*:	The employee is *less* likely to be receptive to his manager's performance improvement suggestions *if*:
He feels his manager is competent in these performance areas and wants to help him to a better job	He feels his manager is incompetent in these performance areas
He is younger than his boss and has had less directly applicable experience	He is older than his boss and has had more directly applicable experience
He was hired or promoted to his present position by his boss	He competed for his boss's position and lost
His work is in good shape and there are no unusual work pressures	He is under unusual pressure at work
He is new in his position or has just been given a new responsibility	His physical or emotional health is not good
He is eager for promotion soon	He is faced with unusual off-the-job pressures
He has just been rewarded with a merit increase or other honor	The manager displays a marked change in attitude toward the employee
Past experience shows the manager will recognize and reward efforts to follow suggestions	Past experience shows the manager has little interest in the employee's response to suggestions

* Reprinted by permission of the publisher from *What to Do About Performance Appraisal* by Marion S. Kellogg. Copyright © 1965 by AMACOM, a division of American Management Associations.

you meet them will in part determine your merit rating and perhaps your next raise. Not only in companies placing this type of emphasis on formal goal statements, but for your own benefit in any organization, it is advisable to set goals with the following in mind.

- Make sure your goals are realistic and attainable.
- Don't set goals casually or make them too simple.
- Leave leeway for unexpected delays or setbacks over which you have no control.
- Reevaluate your goals periodically, and revise them as required to maintain a sensible timetable.

- If a major setback occurs, inform your supervisor so that its effect on divisional and corporate goals can be taken into account.
- Talk over your goals with your supervisor before committing yourself to them.

Whether formalized or not, your ability to set and achieve realistic goals probably represents the major criterion for your advancement within the company.

FACTORS AFFECTING ADVANCEMENT

Quite aside from technical competence, a wide variety of factors can affect your advancement within the company. Many of these depend on your attitude and—let's face it—personality. Selling yourself is important. An intriguing recent article (14) describes two engineers, about equally rated, vying for a promotion. One talked too much and had irritated a lot of people; the other was so quiet that nobody knew him. Who got the advancement? Show your initiative by reading the article to find out!

THE RIGHT PLACE AT THE RIGHT TIME

Being in the right place at the right time sounds like a matter of luck, and it's true that luck can frequently affect your advancement. But there's a lot to be said for searching for the right spot and putting yourself in it. For example, your chances for rapid advancement are small if you enter a well-established organization with a low growth rate, little turnover of manpower or need for new manpower, and which uses specialized technology which must be acquired through on-the-job training. On the other hand, a new company that is rapidly expanding has high manpower needs and may offer excellent advancement opportunities, regardless of your area of specialty. (But the risks of company survival may be correspondingly high and must be taken into account.)

SMALL VERSUS LARGE COMPANY

In a large well-established company, opportunities for advancement can vary widely within the organization. Where there are vigorous development programs involving new products, which may lead to divisional expansion or reorganization, the opportunities are likely to be very good. One of the advantages of a large company is that it is likely to offer a wide range of opportunities within the organization.

On the other hand, you are closer to the action in a small company, and if things go right the advancement opportunities may be tremendous. One thing to remember, though, is that while you may or may not be under more pressure to produce in a small company, you are quite likely to be under closer scrutiny by top management.

TYPE OF JOB

Your job assignment can have a marked influence on the rate of your advancement. The need to acquire specialized skills and technology through on-the-job or special training can delay progress, though it may ultimately lead to better opportunities. Difficulty in judging performance is a related factor. It's hard to assess personal progress on a "blue sky" research project where tangible results may be slow in coming. At the other extreme, some jobs have built-in goals, such as sales quotas for those in marketing. Success here can be measured in dollars, and the figures speak for themselves.

PROMOTION FROM WITHIN

In most companies, it is the policy to promote people from within the organization to supervisory positions and other responsible jobs, rather than to bring in outsiders. This means that you can check the age of people at various scientific and management levels to obtain some idea of how long advancement normally takes. It is comforting, too, to know that you are unlikely to be passed over in favor of an outsider when the opportunity for promotion comes along.

RELATIONS WITH YOUR ASSOCIATES

Since most research and development involves a team approach these days, it is important that you learn how to get along with your associates, from the technicians under you through your peers and up to your supervisor and higher management. While it is not our purpose to instruct you in the fine art of human relations, we will remind you of the advantages of respect for others, giving credit for assistance received, being a good listener, and maintaining a cooperative attitude.

APPLE POLISHING

To deny that "buttering up the boss" has helped some people advance would be naive. But more often it can have the opposite effect and be a great hindrance. It is not a recommended practice. On the other hand, getting to know people—at all levels in the organization—is encouraged. It pays off in knowing whom to contact to get a job done, and in the final analysis it's results that count.

ADVANCED DEGREES

In most companies, coming into the organization with an advanced degree, at the master's or doctor's level, will give you a head start toward advancement, but from then on, it will be results and not any prestige associated with the degree that determines your progress. Many companies do not even allow the use of the title Dr.—everybody is Mr., Miss, or Mrs. (or Ms.) at all levels.

We especially admire employees who have entered the organization direct from high school or military service and perservered to obtain a degree through evening courses. This takes a rare combination of determination and ambition, and these persons are often great assets to their companies.

GRADUATE MANAGEMENT DEGREE

The value to your career of a degree such as the Master of Business Administration depends on many factors. Here we comment on only one of them, its importance to your advancement. Before making the decision for or against going on to an MBA degree, you should get as much information about it as possible, and relate it to your interests and goals. A recent article (15) should be helpful.

While many people feel that an MBA degree may lead to better salaries and faster career advancement, the data (15) are inconclusive. Salary differences are in fact so small that they may well result from variations in working experience, and it is not clear that an MBA really makes it easier to enter management. During the last recession, unemployment among engineers with an MBA degree was higher than among those with an M.S. A large majority of engineers do feel, however, that more formal management training is needed.

Reaction among employers is also varied. Some companies prefer to train their management personnel in their own manner rather than rely on the MBA degree (Fig. 16). A spokesman for a large employer of engineers summed it up this way (16): "In looking at two guys, one of whom has an MBA and, say, both have B.S. degrees in mechanical engineering, we wouldn't lean to the one with the MBA just because he had it, but would appraise both of them for their potential as managers."

"Dave, here, has an MBA--- but we decided to hire him anyway."

FIGURE 16. From *The Schenectady Gazette,* January 10, 1975. Copyright © 1975 by Universal Press Syndicate.

ON-THE-JOB TRAINING

During your first few months in an industrial position, you will almost certainly receive some sort of on-the-job training. It can come in a variety of ways: a tour and seminars on other departments of the company, an orientation course to provide information on the company, courses on specialized technology you may need but weren't likely to get in school (polymer chemistry, for example), or a series of lectures by other staff members.

Some positions, especially those involving close contact with customers, require more intensive training to make you an expert on your company's products. Here you may find a very formal program, with regularly scheduled classes and laboratory sessions, assignments to complete, and examinations to pass. But this time the stakes are higher than just a grade: Your examiners will be customers or potential customers, to whom your opinions and comments are synonymous with those of your company. Learning your lessons well is essential if you are to handle this responsibility adequately.

After your initial on-the-job training, however extensive, it will be your supervisor who assumes major responsibility for guiding your career and assisting you in achieving your goals for advancement. It will be his task to get you to the point where you can carry your own weight in the organization as soon as possible. This is part of his job, overseen and rated by his supervision in turn.

ROUTES FOR ADVANCEMENT

Here is a brief review of some of the commonly encountered routes for advancement in the industrial career:

LINE PROMOTION

The normal route for advancement is by successive promotions through what is known as the line organization into management. In going up the line, one assumes responsibility at consecutively higher levels for larger and larger groups of employees and portions of the operation. This route usually offers more opportunities for advancement than can be found elsewhere. Many scientists and engineers aspire ultimately to management positions achieved by line promotion.

But the corporate structure is of necessity a pyramid, and there are fewer and fewer openings as one progresses, and never enough management jobs to go around. It is for this reason, as well as the personal preferences of some for nonsupervisory duties, that leads many scientists and engineers to find their niches elsewhere, usually in purely technical staff positions.

The urge for recognition and the greater prestige and salary associated with management positions often leads many competent technical people into supervision. Unfortunately, the personal characteristics commensurate with high-quality technical contributions are not necessarily the same ones required to make a good supervisor. If a man's ambition leads him along the wrong path for advancement, both his value to the company and his satisfaction in his job can be diminished.

DUAL LADDERS

Many companies, particularly the large ones, have recognized the problems associated with providing advancement opportunities at all levels for the specialized needs of the professionals. One solution has been the introduction of so-called dual ladder systems, whereby professionals can advance to higher levels of technical responsibility and authority without supervisory duties, as an alternative to line promotion.

A typical dual-ladder system is diagrammed in Fig. 17. Both the names for the various positions and the number of intermediate levels may vary widely from company to company. In principle, positions at the same level are equivalent in status, prestige, and salary. Often one can cross over from one side of the ladder to the other as his goals and quali-

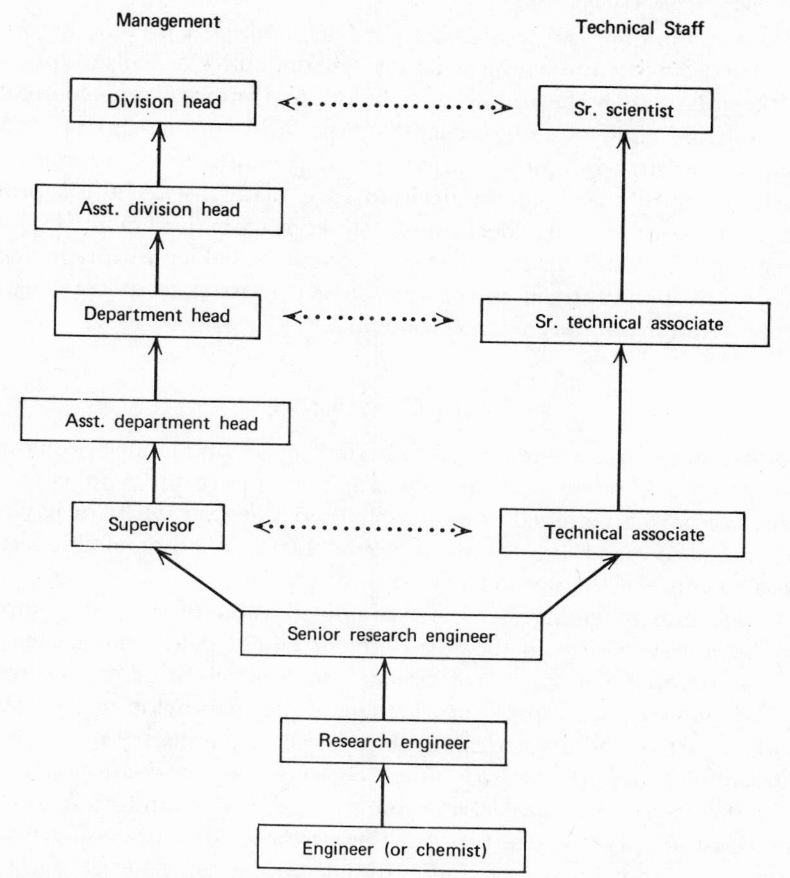

FIGURE 17. A typical dual-ladder structure.

fications, and the available opportunities, allow. A person on the scientific staff usually reports formally to his counterpart at the equivalent level on the managerial staff; as one proceeds up the scientific ladder, his responsibility for independent work increases, and he receives less and less direction from management.

The success and effectiveness of the dual-ladder system depends a great deal on the company's management philosophy (17). At best, it effectively provides opportunities for more high-paying nonmanagerial jobs; at worst, it is an unpopular and unpopulated skeleton around which the competition for line promotion goes on unaffected. In any case, the system is usually confined to the larger chemical companies and some engineering design firms. The small company generally cannot afford many high-priced technical specialists, finding it preferable to buy specialized talent from engineering design companies or consultants.

Other plans than the simple dual ladder are known. Some companies offer a triple choice among generalist, specialist, or managerial routes to advancement, while others provide a dual management approach, in which technical management decisions are made by technical professionals and administrative decisions by supervisory management (18).

Having progressed up the (then new) scientific ladder with a great satisfaction in his industrial career, one of us can attest to its very successful operation in a large chemical company.

CHANGES IN ASSIGNMENT

In addition to, and often as a prelude to, line or dual-ladder promotion, changes of assignment offer a wide range of opportunities for advancement. We have mentioned in earlier chapters the desirability of having a wide range of experience in an industrial career; the rest of this section describes some of the ways to obtain this breadth.

An assignment change within a single division of a large company may be your first step on the advancement ladder. Such changes may be offered to you for a variety of reasons: to broaden your experience in anticipation of later moves into management; to provide manpower to another part of the division in order to meet a temporary need; to test your adaptability to new situations; or simply to motivate you or get you out of a rut. Alternatively, if you are not satisfied and want to make a move to the place where you feel the action is, management is usually receptive to a well-reasoned and justified request for a new assignment.

TRANSFERS

Similar reasoning applies to transfers to another divsion of the company. Again, you may be requested to take a transfer, because your expertise is needed elsewhere, because of manpower needs elsewhere, or even because of the need for a cutback in your division when things get tough. Or, you may have requested the transfer yourself.

Since a transfer of this kind may involve a change of location, you should consider it carefully in terms of both your long-range goals and the attitudes of your family. A factor to be considered carefully is the attitude of your supervision in offering the transfer. Will a refusal have an adverse effect on your future advancement if you stay where you are? You'll need all the advice you can get on this one from those directly concerned.

CHANGING COMPANIES

A statistic that comes as a shock to many new graduates and old timers alike is that these days most new industrial employees change companies after an average of only three years in their first job. Maybe we are old-fashioned, but we think this is deplorable. We suspect that it reflects a poor fit between job and individual from the beginning, and that this stems in part from lack of knowledge of what to expect and how to choose a company and an assignment appropriately. We hope those of you reading this book have better luck.

After three years with a company, you should pretty well know the ropes and have become a valuable part of the organization. To get you to this point has required the company to make a sizable investment; if you leave, this goes down the drain. We feel you owe the company at least the opportunity to try to satisfy your needs by new assignment or transfer, or if this doesn't work out, ample advance notification (one month is standard, but we consider it a bare minimum) so that a replacement can be located.

Without doubt, there will still be many situations where the greater breadth of opportunities for advancement obtainable by changing companies is an overriding consideration. Bear in mind, however, that as you progress into more responsible and higher salaried positions, it becomes more difficult to move to a new company at an even higher level. As you stay with a company longer, retirement benefits and plans for vested interests in them become more and more important. Our personal feeling is that by about age 35 you should be through playing the field, if you have felt the need to do so, and should be ready to settle down to a long career with the company of your choice.

BREADTH VERSUS HEIGHT

Generally speaking, the employees who ultimately reach the top in management are those who have acquired the widest range of experience on the way. Although it is seldom officially admitted, some companies are thought to have an advancement route—so many years here, a period there, a transfer elsewhere—which can be recognized as the training ground for top management. Whether this is so or not, it is interesting to follow the steps in a typical rise to top management. A recent article (19) provides one example.

OTHER FORMS OF RECOGNITION

Up to this point, we have considered advancement largely in terms of higher positions. Obviously, promotions cannot come with high frequency for all industrial employees. Most of us have to be satisfied with other forms of recognition, many of which are at least as beneficial and satisfying, if not more so, than a title on the door and the additional responsibilities it brings.

SALARY INCREASES

Procedures for salary administration and raises vary greatly from one company to another, and you will no doubt have inquired about them during your job interview. Some companies give only merit raises to their professional employees, but in these days of galloping inflation it is usual for cost-of-living raises to be given periodically as well.

Assuming that you are not yet at the supervisory level, you nevertheless are in one of several possible job classifications, as indicated at the bottom of the ladder in Fig. 17. Salary ranges for these classifications usually overlap, but when you get near the top of your classification, your next raise may well be accompanied by a promotion into the next bracket. Because of the overlap, it is conceivable that at times you could be earning more than someone who is in a higher classification. The salary ranges for the various classifications are usually adjusted periodically to compensate for both inflation and increases in starting salaries.

Obviously, the raises that you get will be based on your job performance as evidenced in your performance review. The criteria for an increase vary widely, of course, but the policy of one division of a large chemical company provides a typical example (12). Three factors are considered: the job classification you are in, your experience level and advancement potential, and your performance rating. In this case, the latter is broken down into the difficulty of your goals (weighted 25%), the fraction of these goals that you met (25%), the quality of the results (25%), the timing of the completion (10%), amount of supervisory assistance required (10%), and extra activities (5%).

Since the word usually gets around as to what the average raise for the year is, you can form a pretty good idea of where you stand. You can also get a good feel as to how your salary compares with others in your profession through salary reports published annually in *Chemical and Engineering News*, *Chemical Engineering*, and *Chemical Engineering Progress*. You may also take the (often dismaying) step of comparing your percentage salary increase with the government's figures for cost-of-living increase.

BONUSES AND RELATED PLANS

In many companies, salary raises can be supplemented by other incentives such as profit-sharing plans, bonuses, stock options, and the like.

These fringe benefits can be especially important to employees of a small company, and can really pay off for the outstanding performer.

The award of a bonus may be based on company profits as well as such factors as your length of service, salary level, and value to the company. Some bonuses are given in cash (often the case with relatively new employees), others in company stock. The latter method has the advantage that the amount does not represent taxable income until the stock is disposed of.

A bonus is exactly that, extra recognition for a job well done. It is not to be counted on or considered a part of your regular salary.

PUBLICATIONS

The publication of new research findings and new insights into either technical or nontechnical problems offers you an opportunity to gain recognition both in your company and in the scientific and professional community. The communication of new knowledge and ideas to others is important for both you and your company; it is well recognized that a company's stature and image can be greatly enhanced by the publications of its staff.

Your opportunities to publish will be determined in part by the nature and sensitivity of your job. If you deal closely with proprietary information, you will be less likely to prepare papers for outside publication, but many areas of research and development generate new information of a general and nonproprietary nature which should be published. If you think that your work or ideas warrant publication, and discussion with your associates confirms this, consult your supervision. If you have their approval, you may prepare your manuscript. The guidelines and references for technical writing given in Chap. 4 should be reviewed.

Your completed manuscript must receive the approval of your company. Most companies have well-established procedures for obtaining this approval, involving review by several different departments, in particular the legal or patent group. It is essential that full approval be obtained—and this may require several weeks—and that the manuscript be kept company confidential until the process is completed. These rules apply even if you are writing on scientific subjects only remotely related to your job and company.

TECHNICAL PRESENTATIONS

Presentation of a paper at a meeting of a trade association or professional society is another avenue for obtaining recognition and disseminating new information. In most companies, the manuscript for such a presentation must be written out in full and submitted for company approval by the same route as for publication.

It is often advantageous to present material at a technical meeting before it is submitted for publication in a journal. The discussion which ensues can be very helpful in planning and writing the final manuscript. New information can often be presented more rapidly, and certainly with less formality, through presentation at a technical meeting. There is often a long delay in the publication process, resulting from refereeing, possible revision and resubmission, and getting in line in the journal's backlog. In the more prestigious journals, the waiting time can be up to a year and the rejection rates are high. But don't forget that if your technical society requires a preprint or a written abstract, you must allow ample time for obtaining company approval before the submission deadline. And many companies will not approve an abstract unless they see the full manuscript at the same time.

PATENTS

Patents that you obtain can also be an important factor in your advancement and recognition. This subject is covered in more detail in Chap. 10. Many companies provide substantial rewards, in the form of bonuses or other benefits, to those whose patents are valuable to the company.

CERTIFICATION AND LICENSING

These avenues for obtaining recognition are currently receiving considerable attention. While companies usually encourage employees to pursue certification and licensing, they do not generally require it. These programs relate more to your responsibilities to yourself and your profession, as discussed in Chap. 4.

Many states offer registration and licensing for engineers and chemists. Licensing programs usually require that you pass an examination demonstrating your technical competence. Then you gain the status of, for example, Professional Engineer. You may also become a member of the National Society of Professional Engineers.

Certification programs (20) are offered by some of the professional societies as a means of combatting technical obsolescence by having the proficiency of professionals evaluated periodically. Engineer certification is a process in which an individual's competency to engage in his profession is recognized by his peers. Licensing, in contrast, involves only official legal recognition of competency to practice.

In some states there is now proposed legislation to make the maintenance of professional competence mandatory through continuing education. The American Institute of Chemists and the Society of Manufacturing Engineers already have accreditation and certification programs, and it is likely that many more professional societies will adopt them in the near future. The AIC program requires reaccreditation every five years, through records of courses taken, seminars attended, patents, publications, company reports, and similar evidences of technical competence.

A recent article (21) discusses the controversy over whether chemists should be certified, registered, or licensed.

GETTING INTO MANAGEMENT

If you are like most recent engineering graduates, you probably feel that you would like to get into management eventually. But appealing as this possibility is, it isn't for everyone, and it would be advisable for you to consider carefully whether a management career is what you want to aim for.

Your training in the sciences and engineering sharpens your ability to think through the solutions of technical problems, but by-and-large it does not provide specific training in how to get along with people or to make correct decisions under pressure—two capabilities that are extremely important for effective management. All too often scientists and engineers are not at home when dealing with intangibles, and develop tensions when required to work with people rather than machines. Among the areas you must deal with in assuming supervisory responsibilities and moving into management are human relations, planning, reporting, business organization, accounting, and many others likely quite foreign to your college training. It can be extremely unpleasant to have to undertake responsibilities in these areas, and the long hours of paperwork that go with them, without advance warning and preparation.

The major rewards of management come not from personal accomplishment, but from providing a good environment in which others can produce the necessary technical advances. A manager must receive personal satisfaction from dealing with people, providing the decisions required to meet constant challenges, and participating more broadly in the affairs of his company. The demands, sacrifices, and pressures that come with his job are not justified by the financial rewards alone.

Thus we feel that you should evaluate carefully whether you would fit into management before you decide to pursue it as a career. You should find refs. 22–26 helpful, as well as Chap. 11. Regardless of your long-range plans, it will help your advancement to be aware of the problems and responsibilities of supervision; for this we recommend ref. 27.

REFERENCES

1. Lawrence J. Peter and Raymond Hull, *The Peter Principle,* Morrow, New York, 1969; Bantam Books, New York, 1970.
2. Robert C. Townsend, *Up the Organization,* Knopf, Westminister, Md., 1970; Fawcett World Library, New York, 1971.
3. Shepherd Mead, *How to Succeed in Business Without Really Trying,* Simon and Schuster, New York, 1952.
4. Richard R. Conarroe, *Bravely Bravely in Business,* American Management Association, New York, 1972.
5. Martin R. Miller, *Climbing the Corporate Pyramid,* American Management Association, New York, 1973.
6. Walter S. Wikstrom, "Developing Managerial Competence: Changing Concepts, Emerging Practices," Personal Policy Study No. 189, National Industrial Conference Board, New York, 1964.
7. Thomas L. Whisler and Shirley F. Harper, eds., *Performance Appraisal, Research and Practice,* Holt, Rinehart & Winston, New York, 1962.
8. Robert F. Mager and Peter Pipe, *Analyzing Performance Problems or "You Really Oughta Wanna,"* Fearon Publishers, Belmont, Calif., 1970.
9. Marion S. Kellogg, *What to Do About Performance Appraisal,* American Management Association, New York, 1965.
10. R. B. Finkle and W. S. Jones, *Assessing Corporate Talent, A Key to Managerial Manpower Planning,* Wiley, New York, 1970.
11. P. W. Maloney and J. R. Hinrichs, "A New Tool for Supervisory Self-Development," pp. 343–347 in ref. 7.
12. Herbert Popper, "Performance Appraisal and the Engineer," Part I, *Chem. Eng.* **76** (18), 98–104 (Aug. 25, 1969); Part II, *ibid.,* **76** (19), 133–140 (Sept. 8, 1969).
13. Dennis C. King, "Performance Evaluation," *Chem. Eng.* **78** (7), 110, 112–114 (March 22, 1971).
14. J. M. Vogel, "The Right Men Nobody Wanted," *Chem. Eng.* **80** (26), 242 (Nov. 12, 1973).
15. H. G. Kaufman, "The Graduate Management Degree: Is It Really the Road to Success?" *New Eng.* **3** (2), 29 (February 1974).
16. S. Fisher, "Is There Magic in the MBA Degree?" *New Eng.* **3** (2), 27–28 (February 1974).
17. F. J. Holzapfel, "Multiple Ladders in an Engineering Department," *Chem. Eng.* **74** (5), 124 (Feb. 27, 1967); H. Popper, "Dual Ladders: Thumbs Up or Thumbs Down?" *ibid.,* p. 137.
18. R. R. Ritti, "Dual Management—Does It Work?" *Research Management* **14** (6), 19–26 (Nov. 1971).

19. H. Heltzer, "Until You Succeed You're Stubborn—If You Succeed You're Persistant," *Chemtech.* **1**, 328 (1971).

20. P. Chiarulli, "Certification of Engineers—1, Certification, What It Means," *Chem. Eng. Prog.* **70** (3), 16 (March 1974); C. F. Kay, "2, How the Medical Profession Functions," *ibid.*, p. 19; G. A. Zerlaut, "3, A Certification Program in Being," *ibid.*, p. 23; C. R. Nelson, "4, A Law Against Technical Obsolescence," *ibid.*, p. 28; J. T. Cobb, Jr., "5, What AIChE Is Doing," *ibid.*, p. 31; T. Weaver, "6, The Problem of Continuing Competence," *ibid.*, p. 33; R. Stankey, "7, How One Certification Program Works," *ibid.*, p. 37; Anon., "8, Discussion, Discussion, and More Discussion," *ibid.*, p. 38.

21. Howard J. Sanders, "Do Chemists Need Added Credentials?" *Chem. & Eng. News* **53** (13), 18–27 (March 31, 1975).

22. G. F. Dappert, "Is Managing What You Really Want?" *Chem. Eng.* **74** (5), 120–124 (Feb. 27, 1967).

23. J. J. Smith, "Life of a Manager vs. Death of an Engineer," *Chem. Eng.* **74** (5), 128–131 (Feb. 27, 1967).

24. E. M. Glasscock, "How Others View Your Management Style," *Chem. Eng.* **74** (5), 131–135 (Feb. 27, 1967).

25. A. J. Teller, "Where Will the Creative Engineers Come From?" *Chem. Eng.* **74** (5), 135–136 (Feb. 27, 1967).

26. E. M. Kipp, "Custom-Building Your Management Career, Part I: Establishing Your Performance Record," *Chem. Eng.* **76** (2), 151–158 (January 27, 1969); "Part II: The Novice Manager," *ibid.*, (3), 107–113 (Feb. 10, 1969); "Part III: Section Chiefs and Other Apprentice Managers," *ibid.*, (6) 151–156 (March 24, 1969); "Part IV: "At the Crossroads," *ibid.*, (8) 115–118 (April 21, 1969); "Part V: Middle Management," *ibid.*, (10), 115–120 (May 5, 1969); "Part VI, The Top-Level Technical Manager," *ibid.* (12), 171–177 (June 2, 1969).

27. L. R. Bittel, *What Every Supervisor Should Know,* 3rd ed., McGraw-Hill, New York, 1974.

CHAPTER 6

RESEARCH AND DEVELOPMENT

Before, during, and no doubt for some time after choosing to go into industry, it is only natural for you to wonder whether you have made the right decision. The approach to a new job is always made with many questions unanswered, and some uncertainty and anxiety is inevitable. In our attempts to provide answers, minimize uncertainty, and allay anxiety, we have described the industry, getting the job, responsibilities, and routes to advancement. Now it's time to examine what goes on in the specific job areas to which you may be assigned. Since the odds are better than two to one that your first assignment will be in research and development, it is logical to start there.

We will begin by defining research and development (R&D), then discuss its importance and justification, describe its organization and functions, and point out the groups that support it. Then we will expand upon your role in your first assignment, and some of the problems you may meet. As usual, we will provide the pertinent references for you to dig deeper into each topic if you desire.

THE VALUE, COST, AND JUSTIFICATION OF R&D

WHAT IS R&D?

Scientific research can be defined (1) as the investigation of the relations between cause and effect in natural phenomena. While research can take any of a variety of forms, its main purpose is to generate new scientific knowledge and technology. Development is the application of science and technology to achieve a practical result, such as a product or a process.

Of course, technology can also be advanced by trial and error, without investigation or understanding of the results. This is often referred to as the Edisonian approach, and is discussed later in this chapter. Although this approach is sometimes deplored as inefficient and unproductive, it is nevertheless true that most scientists are guided in part by intuition at some stage of their R&D programs. Certainly, industrial R&D is characterized by both its innovative and its problem-solving natures.

It is commonplace to see such terms as basic, exploratory, fundamental, or applied research, and product or process research and development, as breakdowns of the categories of R&D. To avoid confusion, and despite the fact that there are seldom clear-cut distinctions among these, a few definitions are necessary. When we speak of or provide figures relating to academic and private foundation as well as industrial research, we shall use definitions of the National Science Foundation—NSF—(2). They define *basic research* as "original investigations for the advancement of scientific knowledge that do not have specific commercial objectives, although such investigations may be in fields of present or potential interest to the company." *Applied research*, on the other hand, consists of "investigations that are directed to the discovery of new scientific knowledge and that have specific commercial objectives with respect to products or processes."

These terms, however, do not relate well to the business objectives of industrial corporations, and when we discuss industrial research alone we prefer to adopt definitions recently proposed by the Industrial Research Institute (3). They define three categories: *Research in support of existing business* is, as its name implies, "conducted in direct support of the given companies' existing business" to introduce new products, improve quality, decrease costs, develop new applications, and so on. *Exploratory research* is still in the area of established company interests, but in scouting or preliminary ways. "A new product, process, or service is in view, but the work, by definition, remains 'exploratory research' until a product or process objective is established." Finally, *new high-risk business project research*, sometimes called *new venture research*, "is that conducted with the intention of developing a product, process, or market in which the sponsoring company has no direct manufacturing or market experience, or both." Involving totally new products or processes, it is high-risk in nature.

These categorizations bring to mind the analogy (4) of the R&D process as a long chain stretching between the scientist at one end and the applications man at the other. The latter must get the information from the scientist, but there must not be too much feedback if the research is to remain basic.

HOW MUCH R&D IS DONE?

Here are a few points that emphasize the magnitude and importance of industrial R&D:

- Based on expenditure, about 70% of R&D in the United States has been carried out by industry in recent years (2).
- In 1972, this expenditure for industrial R&D was about $19.4 billion (5).
- This expense represented 3.4% of the net sales of the companies involved (5).
- This amount was split approximately 3% for basic research, 17% for applied research, and 80% for development (5).
- The portion of this expenditure made by basic chemical companies jumped 18% to $1.1 billion in 1974 (6).
- In 1970, there were about 370,000 scientists and engineers directly engaged in industrial R&D (7).
- And the cost of R&D per scientist or engineer in the chemicals and allied products industry was $42,400 in 1972 (7).

Where does the money for R&D come from? Largely, as Table 13 shows, from the federal government. The peak at almost 3% of the gross national product in the mid-sixties was associated with the space program. It is hard to imagine a return to an R&D boom of this magnitude without large government financing of another crash program, perhaps in energy utilization.

While the government finances most of R&D, industry spends the

TABLE 13

Sources of Total* R&D Funding in the United States†
($ billions)

Year	Funds from				GNP	Total as % of GNP
	Federal	Industry	Other	Total		
1961	$ 9.3	$ 4.7	$0.5	$14.5	$ 520	2.80
1963	11.2	5.5	0.7	17.4	591	2.94
1965	13.0	6.5	0.9	20.4	685	2.98
1967	14.4	8.1	1.1	23.6	794	2.97
1969	14.9	10.0	1.3	26.2	930	2.81
1971	15.0	10.9	1.4	27.3	1050	2.60

* Includes funding for industrial, government, and academic R&D.

† Adapted from (8).

TABLE 14

Spending and Source of Industrial R&D Funds in 1971*
($ billions)

Industry grouping	R&D expenditures	Source, % from industry
Aircraft and missiles	$ 4.94	20
Communication and electrical	4.52	49
Fabricated equipment and machinery	2.73	83
Chemicals and allied products	*1.82*	*90*
Transportation	1.76	83
Raw materials	1.40	83
Food and clothing	0.30	80
All others	0.84	—
Total	$18.31	

* Adapted from (7).

lion's share (almost ⅔). Total industrial R&D expenditures for a recent —and typical—year are given in Table 14. It's easy to see what gets the most, and what gets the least, R&D effort, and to which ones the government contributes the most. Overall, the government supports about 42% of industrial R&D, but the defense-oriented categories are the tail that wags the dog. The chemicals and allied products industry is relatively high on the list in terms of dollars, and at the top as far as industrial support of its own R&D is concerned.

Some of these expenditures are further subdivided in Table 15 by type of work, using the NSF definitions for basic and applied research given

TABLE 15

Industrial R&D Expenditures by Type of Work, 1971*
($ billions)

	Percent expenditure on		
Industry grouping	Basic research	Applied research	Development
Aircraft and missiles	1	10	89
Communication and electrical	3	15	82
Chemicals and allied products	*14*	*39*	*47*
All other	3	22	75

* Adapted from (7).

previously. It's clear that the chemicals and allied products industry stands out as putting the most emphasis on basic research; but even so, the fraction spent this way is small compared to that expended in areas more directly aimed at commercial objectives.

THE COST OF R&D

If it is to be justified, R&D must contribute to the achievement of the overall goals of the company in ways that are relevant to the company's current or long-range business interests. The amount of research a company can afford is primarily an economic decision, with larger companies usually able to sustain proportionally larger R&D efforts. The relevance of the R&D program, its organization, and its effectiveness, are all matters of great concern to management. But the benefits of R&D are often intangible, and assessing its contributions to sales, profits, productivity, and economic growth can be extremely difficult.

A common index of the size of an R&D effort, in economic terms, is the percent of net sales represented by the R&D budget. For all industry, this is about 2.3%, but for the chemical industry it is somewhat larger, as indicated in Table 16. Although the average R&D budget for these top ten companies is well over 3% of their net sales, the next ten have an average of only about 1.8%, suggesting that company size does have a lot to do with the amount of research that can be supported.

In addition to percent of net sales, some other common indicators of

TABLE 16

R&D Spending of Top Ten Basic Chemical Companies in 1974*

Company	R&D expenditures ($ Millions)	% of net sales
Du Pont	$325	5.4
Dow	142	2.8
Monsanto	105	3.0
Union Carbide	9.4	1.8
American Cyanamid	57	3.2
Celanese	56	2.8
FMC	50	2.4
Rohm and Haas	42	4.0
Hercules	32	2.0
Allied Chemical	30	1.4

* Adapted from (6).

the size of an R&D budget relate it to profits, value added (that is, increase in value of the product added by the cómpany's efforts), manpower or number of projects in the program, competitor's efforts, and so on.

Deciding how much money should be in the corporate R&D budget is not easy. One usually starts with the current year's amount, and a number of other factors, illustrated in Fig. 18, provide additional input. These are often evaluated in a three-step process (9): "(1) establish a required rate of growth for the company, in terms of either sales, profits, or the share of the market; (2) ascertain what research effort is necessary to develop new products or protect old products sufficiently to achieve the required rate of growth; and (3) establish a research budget which maximizes the probability of achieving the desired research effort and yet minimizes the cost." References 9–12 give more information on budgeting.

Once the total R&D budget has been set by top management, it must be allocated to the specific areas and projects involved. Some companies reverse the procedure, building their budget up from a compilation of the expenditures required for each project for the coming year.

There is no doubt that R&D is expensive, and use of the funds budgeted must be controlled effectively throughout the entire year to prevent significant overruns. This is no small task, regardless of the size of the company. Cost control procedures usually involve dividing expenses into broad categories such as salaries, materials and supplies, equipment, and overhead. (In most cases, at least up to the pilot-plant stage, salaries are by far the largest expense item.) Periodic reports of expenditures are then prepared for each category, providing records of expenditures for comparison to budgeted amounts and planned rates of spending.

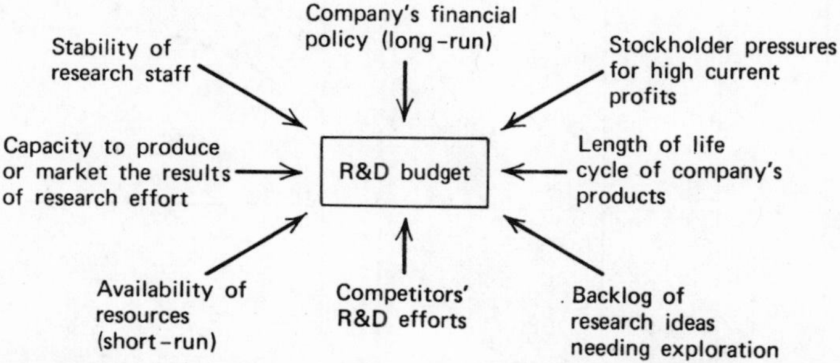

FIGURE 18. Input factors in determining an R&D budget. (From *Improving the Effectiveness of Research and Development* by Robert E. Seiler. Copyright © 1965 by McGraw-Hill. Used with permission of the McGraw-Hill Book Company.)

Inflation, profit squeezes, and other economic factors have caused most companies to tighten control of their R&D spending since the boom era of the 1950s and 1960s. Now more than ever companies are aware of the need for relevance of their research efforts to their business objectives. Industrial R&D must adjust to a seemingly endless list of major challenges, including the balance of trade problem, the changing role of science and technology in our society, the depletion of natural resources and the related shortage of raw materials, the energy crisis, changing R&D incentives, and changing industry–government and industry–academic relations.

R&D AND THE PRODUCT LIFE CYCLE

Every commercial product or process has what may be described as a life cycle, to which we shall refer from time to time in this and following chapters (Fig. 19). R&D plays the predominant role in the early stages of this cycle. Both conception and the establishment of feasibility take place in reseach, and are followed by the development stage prior to commercialization. At this time responsibility for nurturing the product passes to the manufacturing and marketing development organizations described in Chaps. 7 and 8.

Until the product matures (stage 6), there are no significant profits

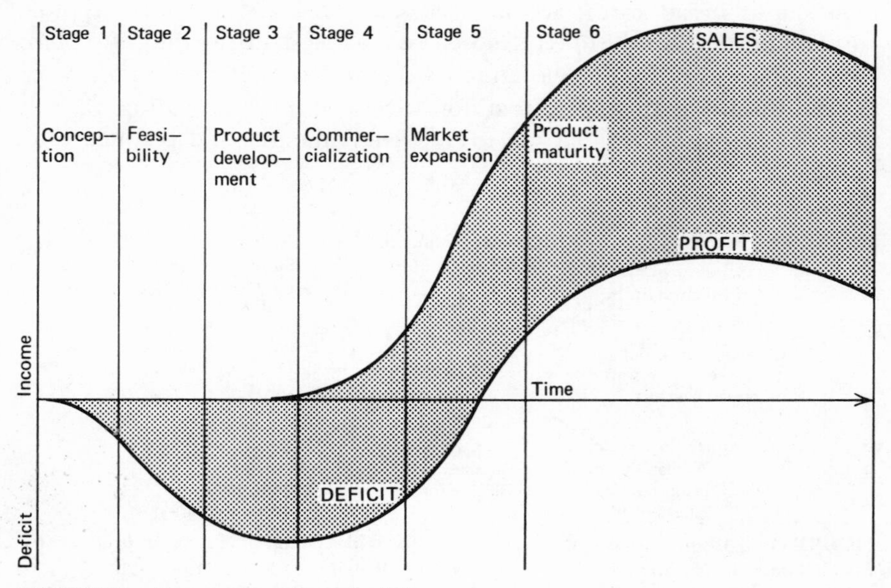

FIGURE 19. The product life cycle (13).

realized; in fact many high cost–high risk projects may never get out of the red. One objective of R&D is to compress the time and reduce the total expenditure prior to commercialization without compromising the quality of the product and the marketing goals set for it.

As the product passes through maturity and competition stiffens, profits begin to decline. This must be combated by improvements in both the nature of the product and the efficiency of the manufacturing processes. A considerable effort by R&D is devoted to improving existing product lines by innovation at this stage.

THE VALUE OF R&D

In addition to exercising control over the cost of R&D, a company must carefully examine the benefits it receives from these expenditures. In some cases these are readily defined, as for example savings resulting from a new, less expensive manufacturing process. But more often it is very difficult to evaluate the economic benefits of the R&D effort. In any case, constant evaluation and examination is required to ensure that short- and long-range R&D programs maintain their effectiveness and their relevance to corporate objectives.

Can the chemical industry afford research? This is the question posed in a recent article (14), which points out that successful R&D is essential for a small company, to provide the innovation necessary for growth. As the company gets big, however, the required return on the R&D investment is so large that a single new product or process cannot begin to justify the cost, and risks are proportionally higher. At this stage, many companies indeed cannot afford to do research; it has become a low-productivity effort that does not pay. Fortunately, a few large companies have developed an R&D organization and philosophy that overcome this obstacle by relying on the net effect of many small projects, no one of which individually has a big impact on earnings.

One such company is Du Pont, as reflected in a recent statement by its (then) president (15): "Asking whether industry ought to do research is like asking whether we ought to keep on breathing. It is not something you do for its own sake, but it is a good idea to keep on doing it if you want to survive. Perhaps some companies have gone into research for the wrong reasons and ought to cut back, but in the high-technology industries the plain fact is that you don't have a choice. Unless you do research you cannot compete."

Simply stated, research is necessary to keep the company alive. At the present rate of increase of technology, a company that does not maintain at least some R&D program directed toward product and process improvement stands a good chance of being left behind. A striking example of what can happen without a timely and properly administered R&D effort is provided by many of the nation's railroad companies at the present time. Innovations and improvements are generally the final products of most industrial R&D programs, and they are vital to the growth of the organization.

WHAT MANAGEMENT EXPECTS FROM R&D

In broad terms, corporate management expects R&D to develop new proprietary products which open new markets for the company, and to improve existing products and processes to keep them profitable. In carrying out these functions, R&D must operate at an overall profit for the company—that is, the expense of R&D must be more than offset by the profits it generates.

In a broader sense, we would like to let the executives of some large companies provide their own answers to what they expect from R&D:

> . . . We look to R&D to keep our product lines competitive! We expect R&D to take part in shaping the future of the company . . . we expect R&D to "anticipate" for the company. [Roger W. Gunder, President, Stauffer Chemical Company (16)]
>
> . . . The priorities for research and development and major problems of a company's top management are inseparable. These priorities are to find ways to get the losing business going in the plus direction, to keep the winners growing, and to find rewarding new areas for new investment. [Robert A. Charpie, President, Cabot Corporation (17)]
>
> The short answer might simply be, "more projects that can be commercialized successfully, and an R&D budget that goes up no faster than earnings." But the long and more accurate answer is more difficult. [Robert S. Ingersoll, Chairman, Borg-Warner Corporation (18)]
>
> . . . To sum it all up, the chief executive really must regard R&D as a profit-center, not a cost-center. R&D managers must think in terms of contribution to profit, not generation of costs. In other words, they must think as managers [John Lobb, President, John C. Lobb Associates (19)]

You may also wish to read refs. 20–23, which discuss the evaluation of R&D relative to corporate profits and objectives.

THE ORGANIZATION OF THE R&D FUNCTION

The organization of research and development groups depends to some extent on the size of the company, and it is convenient to split up the discussion accordingly.

SMALL COMPANIES

Omitting the one-man operations, the organization of a small R&D laboratory typically will follow that illustrated in Fig. 20. Of course, titles may vary and some functions may be omitted, combined, or expanded. For example, one man (or group) may be assigned to support both manufacturing and marketing; or these functions may not be included in the R&D structure. There may be one or more product or process development groups depending on the nature of the company's business. "Long-range" work, if any, includes basic, new venture, and exploratory research. The main point is that in most cases the entire R&D function is centralized in one organization, at a single location. If this location is also the site of the manufacturing operation, there is a strong tendency for R&D to be called on to solve immediate plant problems—the kind of operation known colloquially as "putting out fires." This can be a great distraction to the progress of research.

MEDIUM-SIZED COMPANIES

At this level (Fig. 21), the R&D function is large enough to be housed in a separate laboratory, often at a site removed from the manufacturing plant. There is likely to be a separate plant technical section or laboratory at the plant site, responding to immediate calls for assistance. If the company's product line requires it, there may be a sales service or technical sales laboratory, possibly at a location different from that of the R&D laboratories.

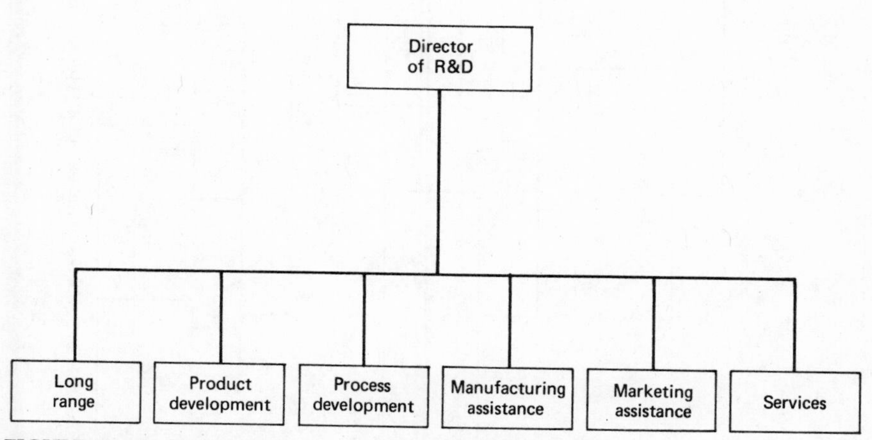

FIGURE 20. Typical organization of a small R&D laboratory.

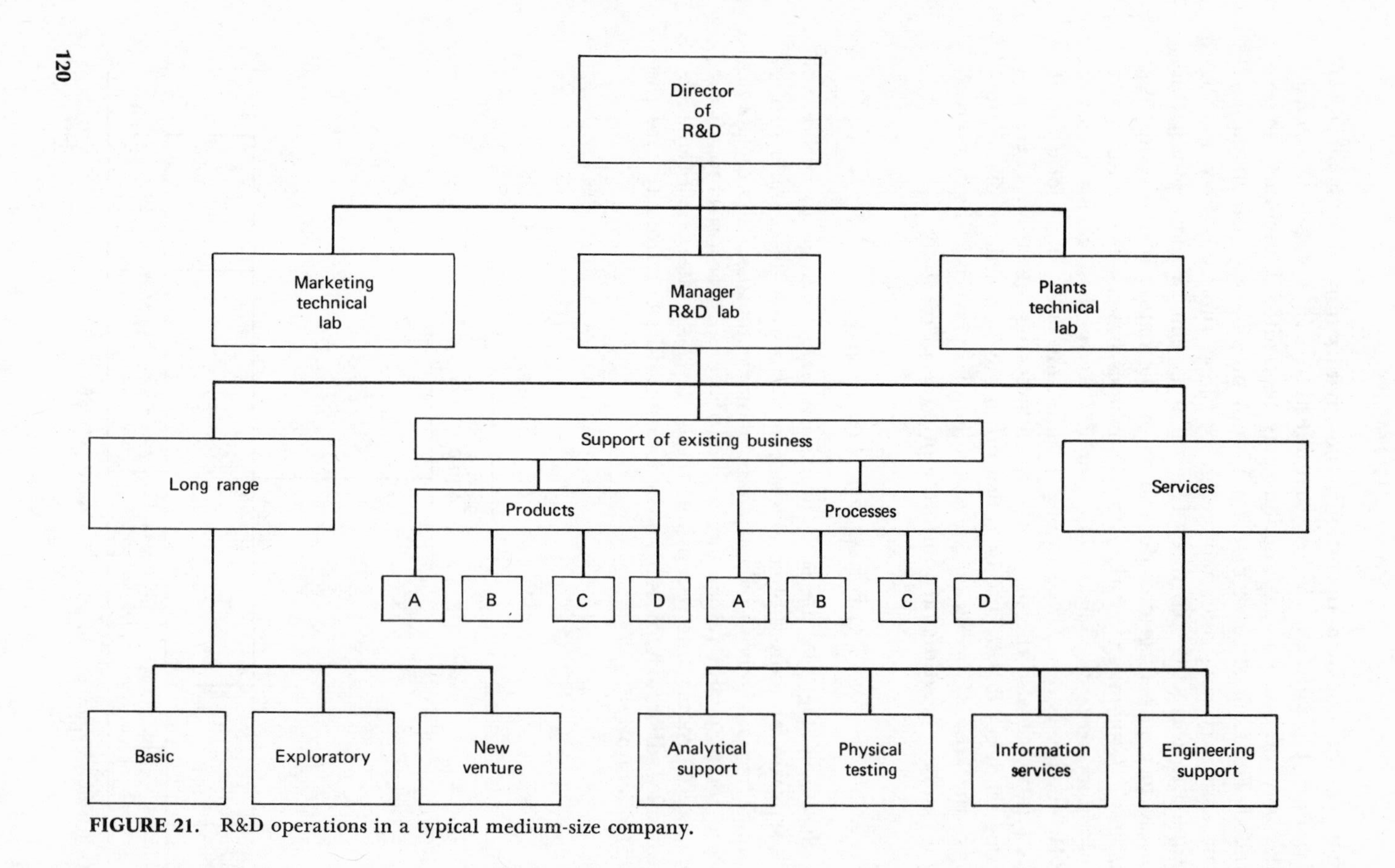

FIGURE 21. R&D operations in a typical medium-size company.

Although the number of groups within the R&D laboratory may have increased, their general character remains the same. A small effort (here labeled long range) is devoted to basic, exploratory, and new venture research. Applied research and development are often lumped together into groups in support of existing business through product and process development—or they may be separate. Service groups are better defined, including analytical service, physical measurements, a library and file room devoted to information retrieval, and a staff providing engineering services. The makeup and responsibilities of these service groups is discussed later in this chapter.

What is designated here as a group may in fact be anything from one man up to half a dozen Ph.D.'s with technicians reporting to them under the supervision of a group leader or supervisor. In some companies, the group leader may be a Ph.D. chemist with technical, but not administrative, responsibility for the research efforts of a small group of M.S. or B.S. holders and technicians. Several groups of this sort may then report to one supervisor who assumes the management responsibility described in Chap. 5.

LARGE, MULTIDIVISIONAL COMPANY

As the company and its R&D division grow to the size where operations are divided among several largely independent divisions, the need to consider the requirements of both the divisions and the corporation as a whole becomes important.

About one in seven multidivisional companies surveyed a few years ago (24) maintain all their R&D in one central corporate organization; in this case the scheme of Fig. 21 applies, but on a larger scale. This arrangement has the advantages that personnel and equipment can be organized and utilized most effectively, and with the least duplication of effort. On the other hand, communication with and response to the R&D needs of the divisions are hampered.

About one-eighth of the companies surveyed had gone to the opposite extreme, having completely decentralized R&D laboratories in each division, structured essentially as in Fig. 21, and no R&D staff at the corporate level. About the same number had R&D executives at the corporate level, but these individuals acted only as advisors to the divisions.

Obviously, advantages and disadvantages of complete decentralization are about the opposite of those of full centralization, and most large companies have adopted the compromise of handling some projects centrally and some divisionally. Their overall organization charts look like those of Fig. 22, with each of the divisional R&D groups organized as in Fig. 21.

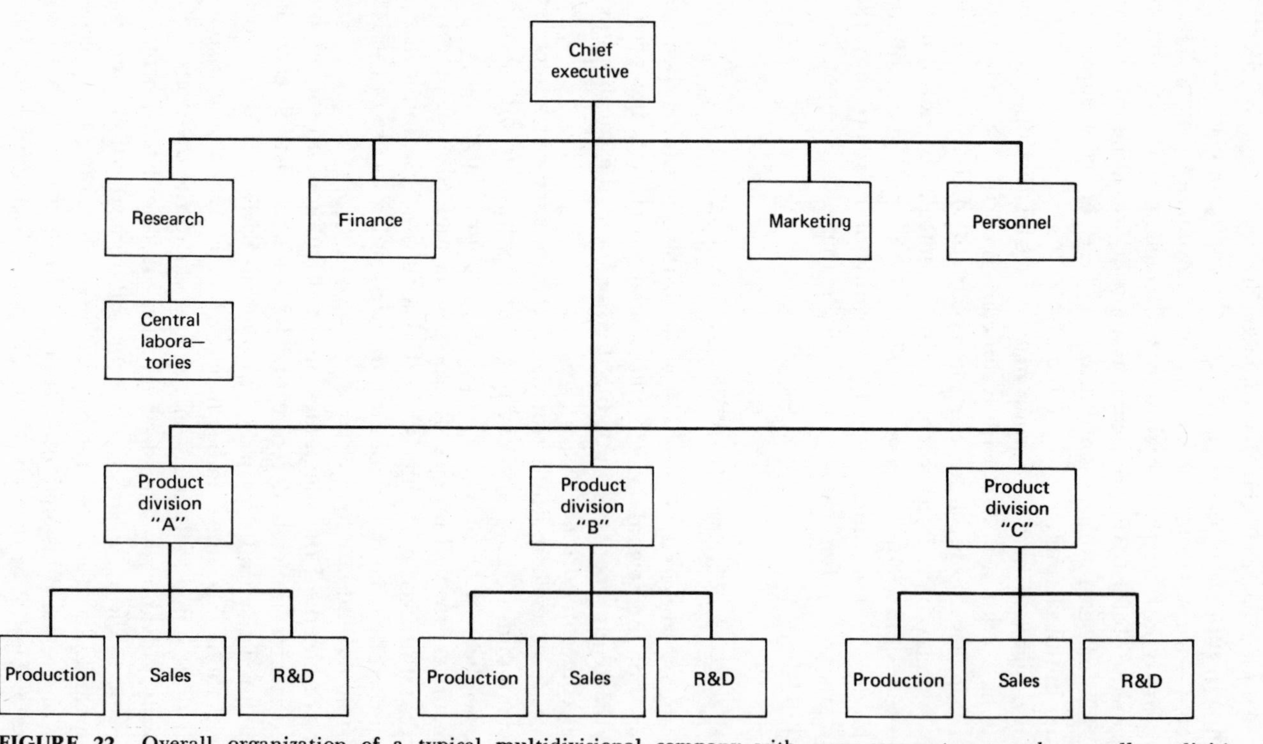

FIGURE 22. Overall organization of a typical multidivisional company with some corporate research as well as divisional R&D (25). (Reprinted by permission of The Conference Board.)

In this arrangement, corporate central research is likely to undertake basic and applied research for the whole company, including work unrelated to existing products or processes, work supporting the manufacturing activities of several divisions, and projects beyond the capability of an individual division because of cost or the need for specialized equipment or specially trained personnel. The divisional R&D groups support their own products and processes and engage in the intermediate and final stages of product and process developments closely allied to their present lines.

An example of this mixture of corporate and divisional research is afforded by the Du Pont Company. Part of all of the R&D effort of each of about a dozen operating departments is located on a central research campus a few miles from company headquarters; no manufacturing is done at the R&D location. A similar campus for sales service laboratories is a few miles away. At the R&D location are the laboratories of two corporate research groups, one concerned with basic research and the other with engineering research. Central facilities for shops, chemical stores, purchasing, and other facilities, and an excellent library and information service provide added benefits from this centrally localized group of laboratories.

TYPES OF R&D

In discussing the various types of R&D, we shall follow the organization of company R&D efforts along the lines of the Industrial Research Institute (IRI) categories of R&D described at the beginning of this chapter, but also attempt to show how the more traditional National Science Foundation categories of basic and applied research fit in. Also we will introduce a few case histories of well-known successful (and sometimes unsuccessful) R&D projects.

EXPLORATORY RESEARCH

The type of research defined by this IRI category is quite similar to the NSF's "basic research": Both are conducted for the purpose of advancing knowledge, both are usually long-range, and both often involve high economic risk. Basic research is usually considered to have no defined commercial objectives, while exploratory research may.

Most basic research is carried out in colleges and universities, as Fig. 23 shows, with the proportion having increased slowly to just over half of total expenditures in recent years. Although the curves in Fig. 23 rise

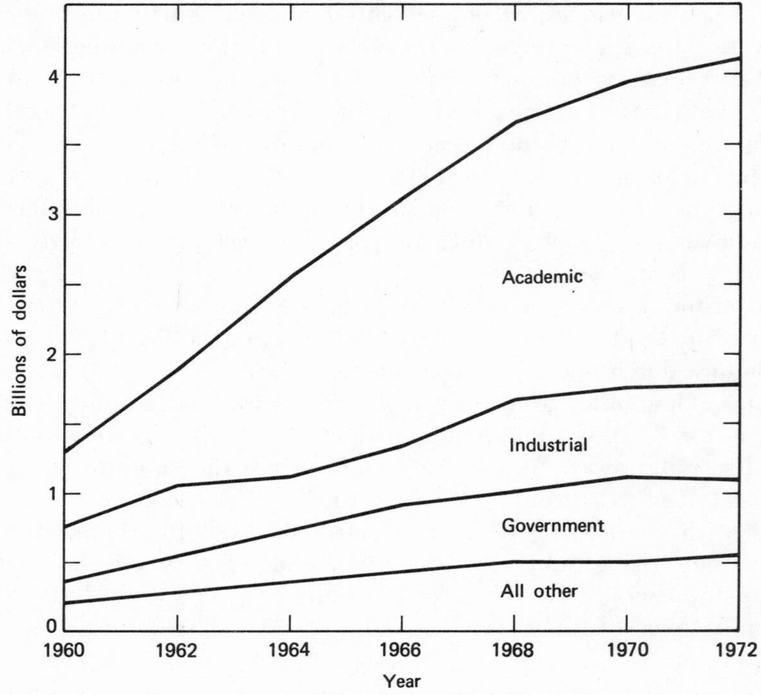

FIGURE 23. Split of basic research expenditures in recent years among colleges and universities, industry, the government, all other research organizations including non-profit organizations and federally funded R&D centers (8).

throughout the time scale shown, conversion of cost figures to "constant dollar" value shows that basic research expenditures reached a peak in 1968 and have declined somewhat since.

Basic research is close to the realization of the classical concept of pure science. "In this country, pure science is pursued largely in university laboratories where the senior scientists pass on the torch to the young students and where the students, through their enthusiasm and inventiveness, help to break down traditional patterns of thought" (26). Whether or not this coincides with your feelings as you make the transition from academia to industry, basic research is a creative intellectual endeavor in which the scientist should be allowed to proceed with minimal constraints. Herein lies the dilemma for industry, since its requirement of making a profit in order to stay alive must necessarily place some restraints on the extent, if not the conduct, of basic research. Table 17 indicates some of the resulting distinctions between basic research in universities and industry. The greatest differences are in the areas of relevance, direction, and communications, as discussed below.

TABLE 17

Basic Research in the University Laboratory and in Industry

Category	In the university	In industry
1. Funding	Mostly external	Mostly internal
2. Personnel	Primarily from academia (graduate students)	From universities
3. Project selection	Unrestricted	Usually must be relevant
4. Publications	Unrestricted	May be restricted
5. Constraints	Funding, personnel, equipment required	Same, but also timing and relevance important
6. Main purposes	New knowledge	New knowledge to generate new ideas which can be commercialized
7. Direction	Should have complete freedom to pursue	Usually closer control and evaluation of progress
8. Supporting groups	Limited	Vast in large company
9. Outside information exchange with experts	Unrestricted	Mostly with consultants
10. Technical judgment	Up to the researcher	Up to the researcher

Most basic industrial research must be relevant to the company's existing business and marketing establishment; it is difficult for any company to justify projects which are not even remotely related to its business interests. This is not to be taken as discouraging basic research, however. Dr. A. M. Bueche, General Electric's vice-president for R&D, recently commented (27): "As long as reasonable balance is maintained, it is absolutely essential for an industrial research enterprise to encourage blue-sky exploration both inside and outside the technical areas of obviout interest to the business. There are numerous examples of blue-sky research leading unexpectedly to new products and services."

Closely related to the question of relevance is the increased supervision and direction present in basic industrial research. Because it is difficult to assess the value of blue-sky research, industrial management appraises its basic research programs frequently to make sure that they are properly balanced with overall corporate goals. Direction, manpower, cost, timing, and results are closely scrutinized. Dr. Bueche says:

> A researcher eventually needs or wants guidance in deciding what kind of work he should be doing. This is no admission of lack of creativity or ability to make major contributions to science. It follows then that the largest share of research people are working not only within the bounds of

> mutually agreeable fields of endeavor, but also under varying degrees of direction. . . . The so-called "directed" people in industry are just as dedicated as the rare self-directed scientists. They have the same curiosity and pride in their work, and they use the same method of discovery that Newton is said to have described to the lady who asked him how he really came to discover the law of gravitation. His reply was: "By thinking about it constantly, Madam."

Despite this greater direction, the industrial scientist has the opportunity to suggest and to work on projects of great interest to him, provided his management agrees. Many companies—Du Pont is one example—encourage their scientists and engineers to bring new ideas to management's attention in the form of research proposals which, if approved, lead to new projects.

The third major difference between university and industrial basic research is the extent of external communication. For obvious business reasons, companies must be cautious about disclosing details of their basic research programs. Thus the flow and exchange of information outside (and in some cases, even inside) the company usually is restricted. The extent of this policy varies widely, and its importance and justification are open to considerable debate, but external communication is almost always less free in industry than in the university.

The NSF (8) recently examined factors impeding the effectiveness of basic research in the United States, regardless of where it is carried out. Many of these were obvious, such as inadequate funding, too few trained researchers, and the like. But among them was inadequate industrial participation: In this view, industry is not doing its share to support basic research. Obvious reasons for this are the almost prohibitively high costs and long time needed to get to the payoffs from this type of effort. Among these payoffs are (28):

- New knowledge, leading to new and improved goods and services—and of course, additional profits to the company.
- A nucleus of in-house experts who serve as internal technical consultants.
- Libraries, analysts, and other support services.
- Inventions which may offer solutions to current problems or provide new applications for products or processes.
- Enhancement of the corporate image through technical papers, talks, and interaction with the scientific community and the general public. The prestige so generated commands public respect and helps attract qualified new employees.

We close this section with two case histories showing the importance of basic research, taken from many in recent articles (26–32) and books (1, 33–37).

Discovery of the Transistor (38)

In 1946 a fundamental research program on semiconductors was initiated at the Bell Telephone Laboratories. This was one of several areas selected for solid-state research. In addition to its intrinsic scientific interest, it was considered an important field because semiconductors at that time already had application in a number of electronic devices such as diodes and thermistors.

Among those working on this project under W. Shockley's direction were John Bardeen, particularly interested in theory, and W. H. Brattain, concerned with several experimental aspects. In 1956 these three received the Nobel prize for the development of the transistor, and this account is taken from Bardeen's Nobel address (38).

The general aim of the program was to obtain as complete an understanding as possible of semiconductor phenomena on the basis of atomic theory. A sound theoretical foundation was available from work on quantum mechanics, theories of photoconductivity, and theories being independently developed on contact rectification.

Research on surface properties was stressed, not only because of the possibility of practical applications, but because it appeared quite promising from the viewpoint of fundamental science. This work, and indeed the entire program, was made possible by the development during World War II of methods of producing extremely high purity germanium and silicon by zone melting, as described later in this chapter.

The basic nature of this research, encouraged by the business interests of Bell Telephone Laboratories in electronic devices which might have applications in the communications industry, but totally devoid of specific product or process objectives, is obvious. So are the payoffs, for the transistor has literally revolutionized all aspects of the communications industry and many others as well. ranging from stereos to computers.

Development of Oral Contraceptives (39, 40)

This case differs from the previous one in many respects. The research must be called exploratory instead of basic, since it was directed toward a very specific product; it was an international effort; it combined the resources of industry and academia, with funding from a nonprofit organization (but not the government); it required crucial management decisions in the face of possible rejection by society; and it succeeded under the impetus of the enthusiasm of a single person.

The foundation of knowledge on which the development of oral contraceptives was based included the results of research on the physiology of reproduction, providing knowledge of the effects of hormones on the reproductive process; on hormonal physiology, providing knowledge of steroid properties; and on steroid chemistry, providing means for the synthesis and structure studies of improved steroids.

The original concepts of oral contraception were untimely. From the first speculations about its feasibility in the early 1920s through developments over the next 25 years, no action could be taken because the available hormones were considered either dangerous or ineffective.

The initial impetus for a research program aimed at a safe, sure, and easily administered contraceptive method came in 1951 from the concern of Margaret Sanger of the Planned Parenthood Federation and Mrs. Stanley McCormick, a philanthropist, over the potential danger of rapid world population growth. In response to their request, Dr. Gregory Pincus of the Worcester Foundation for Experimental Biology proposed a research program to develop an oral contraceptive pill. He subsequently became highly enthusiastic in providing direction for the widespread program.

New synthesis, animal testing, and clinical studies led in just two years to the independent development of two similar agents which were superior in safety and efficacy, by the pharmaceutical companies Syntex S.A. in Mexico and G. D. Searle in this country. By 1954 the time for a large-scale evaluation was at hand. Syntex was not yet in a position to initiate commercialization. Searle, however, made the difficult decision to proceed despite the possible social stigma associated with contraception and the traditional reluctance of an ethical pharmaceutical house to venture beyond the use of drugs to cure disease. The trials, cosponsored by Searle and the Planned Parenthood Federation, were successfully held in Puerto Rico. FDA approved Searle's Enovid for treatment of menstrual disorders in 1957 and for use as an oral contraceptive in 1959. It reached the marketplace in the latter role in 1960. Syntex's similar drug also received FDA approval for menstrual disorders in 1957 and was marketed as Norlutin by Parke-Davis, but that firm would not enter the oral contraceptive market until 1962. Syntex therefore licensed the Ortho division of Johnson and Johnson, who subsequently marketed the product under the trade name Ortho-Novum.

NEW PRODUCT—NEW PROCESS R&D

By far the largest R&D effort in most companies is in *support of existing business*. As defined at the beginning of this chapter, such research can involve new products or processes that are directly related to the company's existing business, existing products (support of marketing), existing processes (support of manufacturing), and possibly even other areas. Because of its size and importance, we discuss the three major aspects of this IRI category in three sections to follow. By the NSF definition, all of this support of existing business research is *applied*, since it is conducted with specific business objectives in mind. In what follows we concentrate more on new products than on new processes.

While it is difficult to generalize because each new product effort has its own specific problems, it is convenient to divide this type of R&D into six phases (41); in a way this is an expansion of the three stages of the product life cycle (Fig. 19) prior to commercialization. They are:

1. concept generation; 2. concept testing; 3. establishing technical feasibility; 4. development of marketing plans; 5. technical development; 6. test marketing.

The concept of a new product may first be recognized in marketing, manufacturing, or R&D. Before it becomes a R&D program, the new concept must be evaluated carefully with respect to technical, economic, and marketing feasibility and relevance to the overall company goals and capabilities. This is by no means easy; launching a major new program in a large company requires extensive planning, justification, and analysis because of the economic risks involved. On the other hand, a small company with a good new idea of smaller scope can often implement the R&D process with minimal evaluation time.

The decision to go ahead with a new product development is made at the top management level on the recommendation of a task group composed of the representatives of the various company divisions involved. The general procedure for the operation of this task group in evaluating the new concept both qualitatively (concept generation) and quantitatively (concept testing), ending in acceptance or rejection by management, is illustrated in Fig. 24.

With approval received, a program coordinator or product champion is appointed and the R&D and corporate venture programs are set in motion. This begins the period of establishing technical feasibility (stage 3) for the product and the process required to make it. Timetables for this stage are set and revised many times as problems are encountered and resolved in the R&D process. It is essential that good communications be established and maintained among the R&D areas involved, manufacturing and marketing.

As soon as technical feasibility has been unequivocally demonstrated, marketing can begin to generate detailed plans for selling the product (stage 4), and at the same time the larger scale technical development work (stage 5) can begin in the manufacturing development groups. The final stage of evaluating the market for the new product usually begins with internal studies, followed with limited outside evaluation just before its official introduction.

Some of the problems associated with new product development, and indications of how they are solved, are listed in Table 18. References 41–49 provide background reading for those interested in or entering this important R&D area. But examples will probably provide the best illustrations of the operation of new product–new process R&D.

The Elastic Fiber (50)

Those in the textile business had long been aware of the commercial opportunity for a durable, dyeable, highly elastic fiber. Since all elastomers as well as all

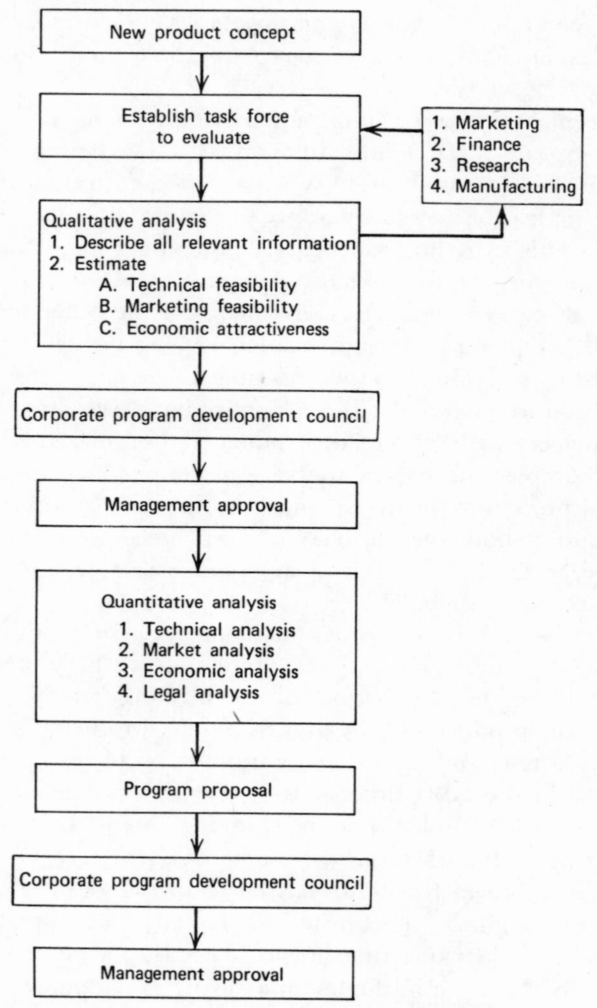

FIGURE 24. Flow chart for launching a new product (41).

fibers are polymeric (51), research to implement this concept could not begin until the field of polymer science had developed. By the middle 1950s enough was known about tailoring the properties of macromolecules through control of their molecular structure to make the attack of this objective feasible. The Du Pont Company through its Textile Fibers Department undertook to do so, and the complicated interplay of research, development, manufacturing, and marketing that resulted is illustrated in Fig. 25.

The key to producing an elastic fiber was to learn how to make a block copolymer—a long chain molecule consisting of alternating stiff and flexible

TABLE 18

New Product Problems and Methods of Solution*

Problem	Suggested solution	Possible methods
Only large organizations are likely to produce high success ratios	Encourage entrepreneurial spirit	Venture teams, product champions
New products don't fit into existing R&D	Organize for new product development	Specific new product group
Organization of R&D is on a functional or project basis	Develop flexible, more responsive project system	Matrix system
New product development must be a continuous process	Utilize a coordinated approach that minimizes slippage	New product coordinating team
Technical and marketing goals conflict	Encourage an active participation of R&D in goal setting	Top-level marketing–technical planning group
A technical breakthrough is needed for success	Identify a real consumer need *first*	Stepwise product development process
Demonstration of technical feasibility is difficult	Conduct concept testing prior to technical development	Same as above
Successful consumer evaluation is needed for success in marketplace	Place the product in an actual selling environment as soon as possible	Controlled store panels, catalog selling
Concept generation cannot be stimulated	Encourage the creative approach within the corporation	Idea generating groups, decentralized and autonomous organization
The new product development process cannot be systematized	Establish specific procedure for new product development	New product group

* Adapted from (41). Reprinted by permission of Research Management.

segments, each made up of many identical repeating chemical units. In the case of the elastic fiber, both types of segments consisted of polyurethanes, the flexible and stiff ones being chemically different. Many polymers of this general nature were synthesized, and the best of them subjected to application research (from left to right at the top of Fig. 25). But the initial product prototype, while having good dyeability and abrasion resistance and twice the elasticity of rubber, yellowed on continued exposure to air. More research was needed, and other polymer segments which did not develop chromophores were synthesized. This

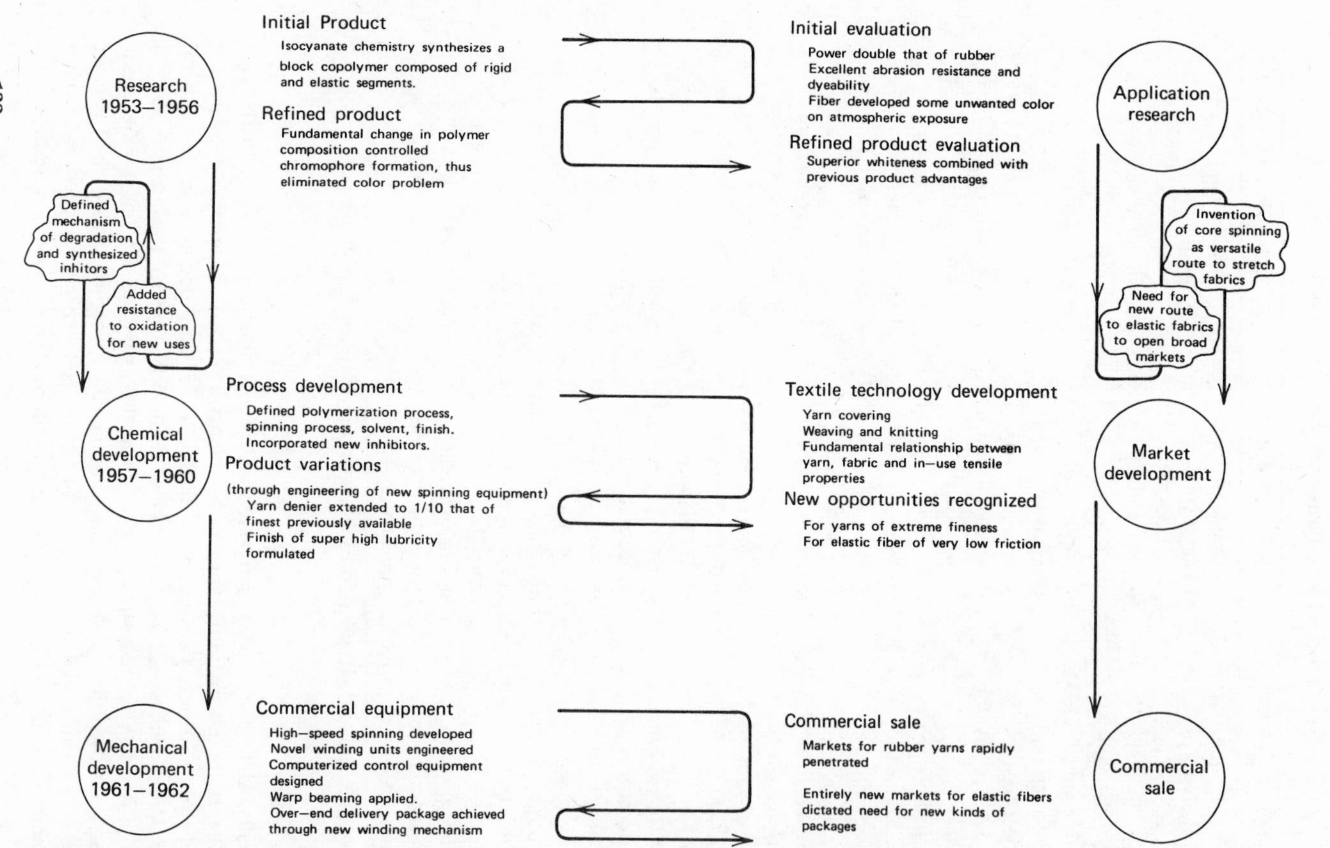

FIGURE 25. Flow chart for the development of "Lycra" spandex (elastomeric) fiber (50). (Reprinted by permission of E. I. du Pont de Nemours & Co., Inc.)

refined product was satisfactory in initial application testing, and both process and marketing development began.

Both of these efforts ran into difficulties that required further research to solve. Further oxidation resistance was needed, and new inhibitors were developed to overcome this problem (proceeding downward on the left of Fig. 25). The proper utilization of the new fiber in fabrics was made possible by the invention of core spinning (proceeding downward on the right of Fig. 25).

With these problems solved, process and marketing development proceeded, always with good communication and close interplay, and evolving over the years into the development of commercial production equipment on the one hand, and test marketing and commercial sales on the other. The resulting fiber was marketed with the trade name "Lycra," and a new generic term, spandex, was coined to describe the class of segmented polyurethane fibers (in Europe, the corresponding generic term is elastomeric). The use of "Lycra" and other spandex fibers for foundation garments, swimsuits, stretch socks, slacks, and sweaters is well-known today.

This R&D effort involved both new products and new processes. It was directed toward a new opportunity within the existing business operations of the company.

Zone Refining (36)

In contrast to the research leading to "Lycra," the zone refining process was developed to meet a specific urgent need associated with the production of the newly invented transistor, described earlier in this chapter.

We saw there that the transistor effect required materials, notably silicon and germanium, of exceptionally high purity. When the commercial usefulness of the transistor began to become apparent, some four years after its invention, it was clear the large-scale production of these semiconductor materials with adequate purity could not be carried out by existing techniques. In direct response to this need, the new and simple technique of zone refining was devised in 1951 at the Bell Telephone Laboratories by W. G. Pfann, a research metallurgist.

In this process, a long bar of, say, fairly pure germanium is placed in a graphite tube. Short lengths of the bar are melted by means of focused heaters, and these molten zones are moved slowly from one end of the bar to the other. Impurities remain in the molten zone as the pure crystalline metal solidifies behind it. Impurity levels can be reduced to about 0.1 part per billion.

The purification of silicon could not be carried out in this way, since silicon melts at over 2500°C and the molten material is very reactive, attacking all known container materials and becoming contaminated in the process. Another invention at Bell, by H. G. Theuerer, provided the solution to this problem. The silicon rod is held vertically and clamped at each end. A short molten zone, held in place by surface tension, is produced by induction heating and moved along the rod. This floating zone technique allows impurity levels to be reduced to less than 0.01 part per billion.

These inventions, stimulated by direct needs arising from the invention of the transistor but tracing back to much earlier basic research in metallurgy at Bell, have had an impressive influence on other fields, with many hundreds of organic compounds and semiconductors, and about a third of all the elements, purified to previously unachieved levels in this way.

"PPO" (Polyphenylene Oxide) Polymer (37)

This case concerns a basic research study carried out in a corporate R&D center, which was transferred to and developed in an R&D laboratory associated with a manufacturing division when a viable commercial objective became obvious.

In 1955, A. S. Hay joined General Electric's Research and Development Center laboratories in the Organic Chemistry Section. In graduate school, Hay had worked on the direct oxidation of organic compounds, and he was encouraged to continue this work at G.E. (a most unusual circumstance, we might add). Hay though he could oxidize phenols by oxygen from the air, using a cuprous salt in pyridine as a catalyst. The reaction would couple the aromatic rings together with oxygen links to form long-chain polymer molecules. Hay was not able to make this work with phenol itself (Fig. 26), but did succeed in making a high molecular weight product through polymerization by oxidation coupling using 2,6-xylenol. This monomer was expensive and not readily available, however, and the polymer had some deficiencies which would obviously require further research to overcome. It was put aside, therefore, in favor of continued research on phenol oxidation.

During this research, G.E.'s Chemical Development Operation (CDO), a manufacturing division R&D laboratory, maintained close liaison with the R&D center. The new polymer held the promise of commercial business, but the CDO laboratory was then fully occupied with the development of another polymer, the polycarbonate "Lexan" (also described in ref. 37).

In 1959, "Lexan" was on a sound footing and the CDO laboratory was ready to undertake another major development. Research on phenol oxidation had not led to polymer products, so Hay's original polymer from 2,6-xylenol was thoroughly reconsidered. In 1960, the two laboratories decided that the business potential was favorable. The key to economic success was the development of an inexpensive synthesis of the monomer, and a team was assigned this task. An answer was found within a year; later a chemical engineer joined this team to develop the commercial process for the monomer synthesis.

Work on the new product gradually transferred to the CDO laboratory as research gave way to development. One key scientist transferred from the R&D center to CDO to continue work on the new product, but most of the staff involved was already at the latter laboratory. Late in 1964, G.E. announced the commercialization of "PPO," and pilot-plant quantities of the polymer became available early in 1965. Up to this time, G.E.'s total investment in R&D on the new product was $7 million. This was the depth of the "valley" at the end of stage 3 in the product life cycle chart of Fig. 19. As we shall see in later case histories, this was not an unusually large sum.

"PPO" is a strong, durable engineering plastic with excellent dimensional

"What's the opposite of 'Eureka!'?"

FIGURE 26. From *The New Yorker* **50**(47), 32 (January 13, 1975). Drawing by Dana Fradon. Copyright © 1975 by The New Yorker Magazine, Inc.)

stability at elevated temperatures, good electrical properties, and good resistance to water absorption.

GS Hydrophobic Silica (52)

The final case history in this group is taken from one of the most engaging little books describing a career in industrial chemistry we have ever seen. It is highly recommended reading.

Guy Alexander joined the Du Pont Company in 1947 to work on silicate chemistry in a group supervised by Ralph Iler. One of Du Pont's products at this time was "Ludox," a colloidal silica sol used as a thickening agent and to produce nonskid properties and other improvements in floor waxes.

One day Iler told Alexander about a trip report from a salesman indicating a need in the petroleum industry for greases that do not become fluid at high temperatures. Most greases are made by thickening oil with soap; at elevated temperatures, the soap melts and the grease gets too fluid to be effective. Could the silica powders made by drying "Ludox"' be used to thicken oil? Being inorganic, they would not melt as the soap does.

It was first found that the efficiency of a silica powder in thickening oil was an inverse function of its bulk density, and efforts were made to produce lighter, fluffier powders. The percent silica needed to thicken a given lubricating oil was reduced in this way from 30 to 50 to around 10. The silicas, however, were hydrophilic because of silanol groups on their surfaces (Fig. 27*a*). In the presence of water, the surface became hydrated (Fig. 27*b*) and the silica was extracted from the grease. This instability could be overcome if the surface could be made hydrophobic.

It was known that treatment with trimethylchlorosilane would produce the desired effect (Fig. 27*c*), but a simple estimate showed that this would cost about 40¢ per pound of silica, whereas the silica would have to sell for around 50¢ to be competitive with other grease thickeners. Thus, a less expensive treatment was required, and this turned out to be esterification with butanol (Fig. 27*d*). The product was satisfactory, and was named GS Hydrophobic Silica.

Soon a team of two chemists and an engineer was working with Alexander on the development of the new product. As further difficulties were uncovered and solved, the engineering aspects of the problem became more important, and Alexander passed the responsibility on to others in order to remain in exploratory research.

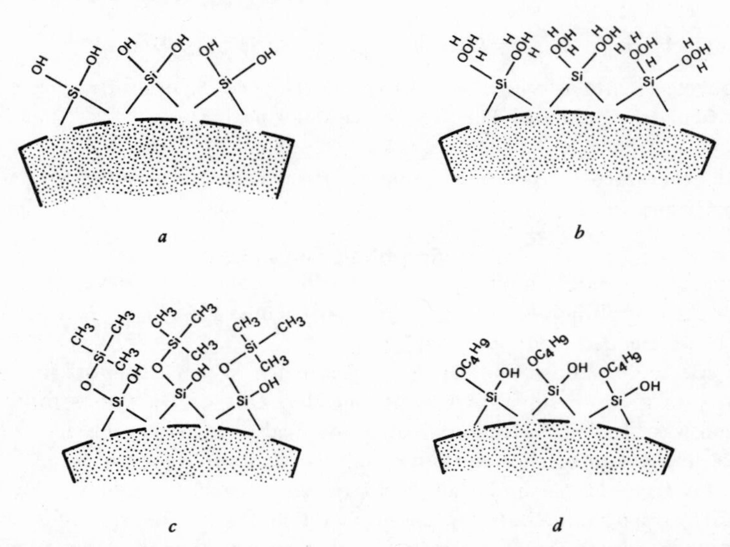

FIGURE 27. Schematic diagrams of a portion of the surface of a silica particle about 5 nm in diameter (52): *a*, normal silanol surface; *b*, hydrated surface; *c*, a hydrophobic surface obtained by treatment with trimethylchlorosilane; *d*, esterified hydrophobic surface of GS Hydrophobic Silica. (Reprinted by permission of Dr. Guy Alexander.)

Although they differ rather widely, these case histories show a number of similarities characteristic of most new product–new process R&D in support of existing business:

- The research is generally motivated by the combination of a need and a technical opportunity, characteristic of successful innovation. Individual creativity plays an important role.
- The new product is usually initiated in research, where its technical feasibility is established. The new technology is then progressively transferred to development, engineering and supporting functions, and finally to manufacturing.
- The successful commercialization of a new product involves almost every area of the company. It requires cooperation, good communications, and frequent and accurate reviews of technical, economic, marketing, and legal aspects relative to corporate goals. The mortality rate is high: Only one in four new products introduced to the market is successful.
- Because of their relation to existing business, the examples described do not represent high-risk, new venture programs.

RESEARCH IN SUPPORT OF MARKETING—EXISTING PRODUCTS

When a product has reached maturity (the sixth stage in its life cycle, Fig. 19), improvement is essential if obsolescence is to be staved off. Examples of product improvement R&D are common in everyday life—fluoride toothpaste, belted bias and radial tires, wash-and-wear fabrics, faster and better photographic films, and many more. The incentive for research in support of the marketing of existing products can arise in many ways:

- Perhaps a customer is having difficulty with your product; the field sales representative reports it to the home office, and marketing requests R&D to find out what the problem is and how to solve it.
- Perhaps a competitor has improved his product, putting yours at a disadvantage. To regain your competitive position will require significant product modifications, and an R&D program—which may prove very costly—is requested.
- If approved, pending legislation will have a significant impact on the sale of your product because of potential environmental problems. A substitute meeting the requirements imposed by the legislation will have to be developed if the business is to be preserved.
- Exploratory research has uncovered new knowledge which can lead to an improved product, increased sales, and advantages to the customer.
- Your product could meet the requirements of a new application if it

were improved in a certain property, for example, high-temperature stability. It appears technically feasible, and the increased sales volume would justify the R&D expenditures.

The routes by which a product improvement program is conceived and developed are quite similar to those discussed earlier for new products, and we won't repeat them here in detail. The concept can arise almost anywhere along the chain of research, development, manufacturing, and marketing. Much more important than where it arises is the need for effective communications all along this chain if the needs and opportunities for product improvement R&D are to be recognized and proper action taken. Since the customers reap the ultimate benefits of product improvement, a good market research program is essential to assess their requirements and reactions.

It certainly makes good business sense to pursue a balanced program of R&D on the improvement of both products and processes. In fact, we have felt for some time, and a recent article (53) expresses the same view, that the development of major new products and processes will probably be less important in the decade or two to come than it was in the 1960s in determining a company's growth. Rather, strategy will be aimed more at the improvement of existing products and processes.

RESEARCH IN SUPPORT OF MANUFACTURING—EXISTING PROCESSES (54)

Many of the traditional incentives to improve manufacturing processes are obvious, including new knowledge of important variables, new technology outdating the present process, product improvements requiring process changes, business conditions making improved efficiency essential, and many more. Today these reasons for existing-process R&D are supplemented—and often outweighed—by a group of compelling new factors including environmental legislation, OSHA safety requirements, the unavailability of many raw materials, and the lack of sufficient energy at reasonable cost.

Whatever the reasons, there is no doubt that process improvement is a significant route to increased economy in manufacture, leading to both decreased selling price and increased profits. The total impact of R&D on product and process improvement in the Du Pont Company is illustrated (53) by these changes in selling price and production volume between 1960 and 1971: for "Dacron" polyester fiber, staple and tow, price down 67%, sales up 700%; for nylon BCF fiber, price down 43%, sales up 2000%; and for "Mylar" polyester film, price down 38% and sales up 300%.

Because of its close association with manufacturing, most of the work aimed at improvement of existing production processes falls in the category of development rather than research. Development bridges the gap between research and manufacturing, playing important roles in technology transfer, in programs for improved products as well as processes, and in troubleshooting or problem solving for manufacturing. When production problems occur, it is usually up to the development or plant technical staff to provide the technical solutions.

Even more than in the other areas of R&D, economic considerations strongly control the development of improved processes. The company already has considerable capital invested in the equipment and machinery for the existing process. If the improvements are compatible with the existing hardware, their cost will clearly be less than if new equipment or radical changes are proposed. These must be justified economically in terms of savings in manufacturing costs and product improvement. It goes without saying, however, that the existing process and equipment should not stifle innovation or preclude consideration of all possibilities by the process R&D groups.

PROBLEM SOLVING

It should be obvious that a significant portion of R&D, and particularly of the development engineer's effort, can be characterized as problem solving. We pointed out in Chap. 3 that industry is continually looking for employment candidates who have demonstrated the ability to tackle and solve problems. It seems worthwhile to digress a moment to examine a few of the guidelines for problem solving that are particularly suitable for application to R&D (55*):

Run over the elements of the problem in rapid succession several times, until a pattern emerges that encompasses all elements simultaneously.

Suspend judgment. Don't jump to conclusions.

Explore the environment. Vary the temporal and spatial arrangement of the materials.

Produce a second solution after the first.

Critically evaluate your own ideas. Constructively evaluate those of others.

When stuck, change your model of the system. If a concrete representation isn't working, try an abstract one and vice versa.

Take a break when you are stuck.

Talk about your problem with someone.

NEW HIGH-RISK BUSINESS PROJECT RESEARCH

As the IRI definition cited early in this chapter states, the distinguishing characteristic of this type of "new venture" research is that it takes the company into areas in which it has no previous direct manufacturing or marketing experience. R&D programs of this type are usually carried out to achieve diversification; that is, the creation of new marketing areas for the company.

It is also characteristic of this type of R&D that the risks of failure are higher than those incurred in following the company's traditional product lines. If the research is technically sound, the development of product and process can usually be carried out successfully, but the essential element of feedback from the marketplace is missing, and the acceptance of the product, ensuring a viable level of sales, is gravely difficult to predict. A famous case history serves as an example.

The Corfam Saga (50, 56)

For many years the Fabrics and Finishes Department of the Du Pont Company made coated fabrics, in which a thin layer of polymer in film form was laid on and bonded to an underlying fabric. It was tempting to consider these products as substitutes for shoe leather. Indeed they could be made with the proper toughness and flexibility for that application, but they failed utterly because of their complete lack of permeability to water vapor.

Here was a technical challenge that Du Pont thought it could meet. It was without doubt a high-risk business project, not only because no one had ever made or knew how to make a leather substitute, which might or might not bear any similarity to the coated fabrics with which the F&F Department was familiar, but particularly because its sale would require Du Pont to penetrate entirely new marketing areas. In keeping with general company policy not to enter directly into producing consumer products, Du Pont did not intend to make shoes, but only to sell a leather substitute to shoe manufacturers.

In the 1950s, Du Pont assembled a task force of over 50 scientists and engineers to attack the problem of a material with permeability to water vapor, but not liquid water, with other properties appropriate for a leather substitute. This is said to have been one of the company's most difficult R&D projects, and the total development cost tens of millons of dollars.

The result was a technical success. Corfam poromeric material (56) was a multilayer structure in the form of a nonwoven polyester web fabric, which was punctured with needles, producing millions of microscopic holes per square inch, to provide vapor transmission, and then impregnated with a polyurethane elastomeric binder. This two-layer microporous structure was then covered with a thin vapor-permeable polymer coating to obtain a smooth, grained, or suede surface, and often there were other intermediate woven fabric interlayers.

The process for making Corfam by the steps of forming the fibrous web, impregnating it with polymer, providing porosity, coating, and finishing, was exceedingly complex, once comprising 20 separate manufacturing steps, and much new equipment of types that had never been built before.

Corfam met its technical goals with plenty to spare. It duplicated the natural combination of properties desired, and added product uniformity not found in leather. No shoe material lasted longer or held a shine better. But Corfam did not behave like leather. For one thing, it did not stretch, so that one could not buy a pair of Corfam shoes a little on the snug side and expect them to "break in."

For a variety of reasons, this truly superior material never achieved widespread public acceptance. A technical success, it was a financial debacle. The relatively high cost of the product compared with nonpermeable vinyl shoe uppers—admittedly inferior from the standpoint of comfort—priced it out of the market. In 1971, Du Pont announced (57) that within a year it would no longer manufacture and sell Corfam. Du Pont spent between $80 and $100 million on this high-risk business venture over the 7-year life of the ill-fated product. Inventory on hand and the trade name were sold for about $6 million to a U.S. leather goods supplier, and the manufacturing equipment and rights to sell (except in the United States and Japan) were sold to a Polish, state-owned manufacturing company.

SUPPORT GROUPS WITHIN R&D

In your university years, you no doubt made use of one or two small support groups or facilities within your department. There was probably a machine shop (perhaps shared among several departments), the library, the stockroom, and a glassblower for the chemists. Though you may not have been directly aware of them, the university also had service facilities such as shipping and receiving, purchasing, maintenance. And, of course, no self-respecting school would be without its computer facility these days.

Industry has all these facilities as well, and some of them are described in more detail in Chap. 9. Here we wish to bring to your attention some specialized support groups, not likely to be found in universities, but common to most good-size industrial R&D organizations. As listed in Fig. 21, they include analytical support, physical testing facilities, information services, and engineering support. These and other areas will be discussed in relation to your first assignment later in this chapter.

ANALYTICAL SUPPORT

In industry, analysis is a very specialized profession. Whereas in the university you probably did your own analysis, if indeed any were done, in industry you will find a group of experts ready to perform these tasks for you. It's their responsibility to see that these jobs get done right, developing new methods and equipment if needed to provide the necessary information.

The modern industrial analytical group is heavily oriented toward instrumental methods, because of their versatility, better precision and accuracy, and the better utilization of skilled manpower that they provide. It is generally accepted that the investment in good analytical instrumentation pays for itself many times over. In fact, some of the best analytical instruments have been developed in industrial R&D laboratories.

Although company interests and the nature of the product line determine what analyses are important, a good analytical group in a large R&D division will be prepared to perform any of a wide variety of analyses—spectroscopic, from X-ray to ultraviolet, visible (including emission and atomic absorption), infrared, NMR, and EPR spectroscopy, mass spectrometry, gas and liquid chromatography, electrochemical and radiochemical analyses, and all sorts of optical and electron microscopy. The general rule is you ask for it, they'll do it.

But don't abuse this privilege. Analyses take time and cost money. If you are a smart scientist or engineer, you won't ask for "the works," but use your analytical service group as consultants. Take your problems to them; ask them what analyses they would recommend. On the basis of their experience, they will know what can and can't be done. Consult them also on the interpretation of the results—what they mean and how reliable they are. If you don't keep in good communication with these experts, you aren't using their services properly. These remarks, in fact, apply to each of the service groups discussed in this section.

PHYSICAL TESTING FACILITIES

Many chemical products, including metals, polymers, and ceramics, are materials which are sold largely on the basis of their physical, rather than chemical, properties. The determination of the physical, optical, mechanical, and electrical properties of materials is not unlike their analysis for chemical composition in concept, but the tools and techniques are so different that these measurements are usually made by a separate support group.

Some of the mechanical properties that may be measured by a physical testing group are tensile, shear, flexural, and compression strength, elongation at break, stiffness, fatigue properties, impact strength, abrasion and tear resistance, hardness, and such thermal properties as softening or melting point and flammability. Optical tests include transmittance and reflectance, color, gloss, haze, transparency, refractive index, and the like; electrical tests, dielectric constant and loss factor, dielectric strength, resistivity, and arc resistance. Among the other properties that may be important to test are chemical and solvent resistance, vapor permeability, and resistance to a variety of environmental factors such as outdoor exposure. Many of these tests must be carried out at high and low temperatures as well as room temperature.

INFORMATION SERVICES

A good industrial information services group will go far beyond the already wide variety of services offered by a library. Typically, one will find files of company reports and documents, indexed in a variety of ways for rapid information retrieval; facilities for literature and patent searches and translations; and in the larger companies (such as Du Pont and Kodak), literature and patent abstract bulletins not unlike *Chemical Abstracts* in format—all these in addition to current and retained journals, books, and patent files. Don't overlook the importance of using these facilities regularly, aside from your immediate job assignment, in order to combat technical obsolescence.

ENGINEERING SERVICES

Many R&D divisions include a group consisting, typically, of chemical, mechanical, electrical, and environmental engineers and experts on instrumentation. It is the responsibility of this group to design and build at the semiworks and plant-scale level. Here attention must be paid to such processes as heat transfer, fluid flow, mixing, pollution control, and a variety of what the chemical engineer likes to call unit operations, which the bench chemist can largely ignore because of the small scale of his experimental setups. If your R&D division doesn't have this type of engineering services group, it may be accustomed to purchasing these services on the outside. (Indeed, all of the services described in this section can be so purchased; at what point it is desirable to do so is discussed in the next section.)

SPIN-OFFS

The maintenance of these service organizations is expensive, needless to say. Since the extent to which they are used is likely to vary with the nature and size of the entire R&D operation, the problems of justifying the investment in service equipment and personnel and of keeping an adequate force to meet emergencies must be given careful consideration. In 1972, two major companies (Du Pont's Plastics Department and Dow) announced (58) that they would offer the sale of analytical services, primarily in the form of advanced testing involving instrumentation not available to the small analytical laboratory. Although the public announcements do not stress it, the advantage to the company in keeping its staff together and operating in slack times is at least as important as the additional revenue from this venture.

Another type of spin-off from sophisticated R&D can take place when an invention or development is made which is outside the company's field of direct interest, and perhaps too small to warrant a new venture for a large company. For example, research on radiation damage of "Lexan" polycarbonate resin at the General Electric Company led to the development of uniform microporous filters. A new small company was started with G.E.'s help to manufacture these filters under the tradename "Nuclepore." And in 1971 Dow sold an entire 36 million lb/yr phenol plant to three former employees. Their new company was to sell the product back to Dow. Finally, the analytical and engineering divisions within Du Pont developed many instruments for process control as well as analysis over the years. The manufacturing rights for these were often sold to small instrument companies, who would produce the units for Du Pont or for sale to the public. Even after Du Pont formed an Instrument Products Division which develops and sells instruments on its own as well as in cooperation with other chemical groups, this practice continued.

YOUR FIRST ASSIGNMENT

There's no doubt that life in industry is vastly different from that in the university. Some of these differences have been alluded to in previous chapters since they affect your preparation for, search for, and new responsibilities in an industrial job. Now we wish to provide what we hope will be useful tips as you are ready to start that first job assignment in R&D.

BE READY TO SHIFT GEARS

Don't be unaware of the differences between academic and industrial R&D. If you look back at Table 17, you'll see that most of the differences lie in the areas of economics, relevance, direction, and the greater emphasis on applied R&D in industry. At the university, your research goal was to advance the frontiers of science through the creation of new knowledge; this was basic or fundamental research. By the very nature of industry, you will be applying new knowledge to provide saleable goods and services and to make a profit for your company.

It goes without saying that your ability to think is a prerequisite to success in both academic and industrial research. But in industry the ability must be combined with rapid and correct application of that thinking. There's no question that time is money in industry; the pace is rapid, the stakes are high, the rewards are significant—but the pressures to produce can build up, too.

While the rules of the game are different, one reward remains much the same in industry as in the university. That is the personal satisfaction which derives from new discoveries, problem solving, teamwork, achieving goals, and the respect of your associates.

MAKE APPROPRIATE USE OF YOUR NONTECHNICAL ASSISTANTS

In most R&D laboratories, you will enjoy a luxury not found in the university, the help of a technician or laboratory assistant. The amount of help you have will depend on the nature and priority of your work, but typically you can count on having someone available to assist you about half the time. If you plan for and use this help effectively, it can be invaluable to the smooth progress of your research.

Some of the tasks commonly assigned to a laboratory assistant are preparing routine experiment setups, carrying out simple reactions or syntheses, cleaning up, drawing charts for presentations, and many more. We emphasize that it should not be beneath your dignity as a professional to do these things if need arises, but it makes more sense to have them done by your assistant.

It is very important to remember that with the responsibility for assigning a job to your assistant, you assume responsibility for his safety on that job. You must be absolutely sure to inform him of any possible hazards, and to see that he understands fully all the necessary precautions and procedures to carry out your instructions safely.

USE THE SUPPORT FACILITIES PROVIDED FOR YOU

High on your list of things to do in your first assignment should be to get acquainted with the wide variety of support you can call on for help in tackling your problem effectively. Most of these support areas (going far beyond the group specific to R&D divisions mentioned in the previous section) are shown in Fig. 28. Some of these, such as a large technical

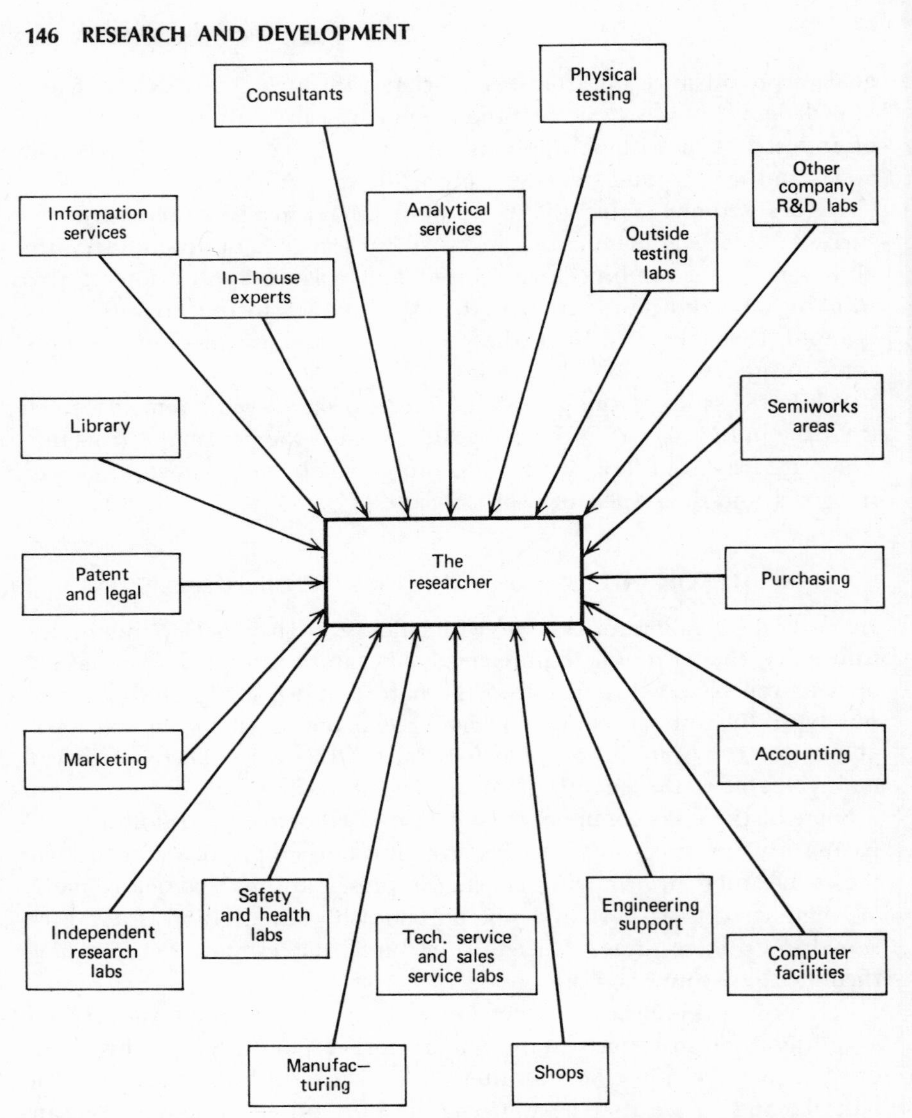

FIGURE 28. A few of the types of support available to the industrial researcher.

library or computer facility, may be too expensive for a small company to maintain, but these services are available from outside organizations.

Proceeding clockwise around Fig. 28 from 12:00, we note that the *analytical and physical testing services* already described are basic for the evaluation of research results. If the test facilities required are not available within the company, the services of *outside testing laboratories* can be used, as we mentioned (58). Before submitting requests for testing to an outside laboratory, you should discuss the project with your supervisor, and in particular make sure through the legal department that no proprietary information is divulged. (If necessary, the service laboratory can be asked to sign a confidentiality agreement.) The major incentive for taking work outside should be research efficiency: If the job can be done better, faster, and more cheaply outside, that is where it should be done. On the other hand, don't overlook the fact that, in a large company, *other R&D laboratories* may have the necessary facilities to develop or carry out specialized tests.

When the time comes to think about the commercial possibilities of a product or process, you can make use of the facilities of a *semiworks area*, with equipment allowing the duplication of the production process on a smaller scale. In addition, these facilities can be used to provide the quantities of materials needed for large-scale evaluation programs. If they are not available within your company, semiworks services can be procured from external sources.

In a small company, you may be "on your own" in locating sources for these outside services as well as for chemicals and equipment. In a large company, you can take advantage of the facilities of your *purchasing department* to locate these sources and obtain prices and delivery information. Likewise, the *accounting division* can be of great assistance to you in evaluating the economics of a product or process, as well as keeping track of your research expenditures. These and many other divisions of the company which can help you in your R&D activities are discussed further in Chap. 9.

The *computer facility*, either in-house or outside through time sharing, is also there to help you in the mathematical, statistical, and data-handling aspects of your research. Although in the university you may have been encouraged to do your own programming, in part to gain familiarity with the power and limitations of computers, we encourage you to work through professional programmers when you are in industry. It's a matter of efficiency again: They can usually do the job quicker and better.

As discussed earlier, *engineering support* facilities can assist you in the areas of the design and construction of new equipment, setting up semiworks or pilot-plant scale operations, and establishing engineering specifications and requirements for new production processes. On a smaller scale, a variety of *shop facilities* is available for construction and custom fabrication in glass, metal, wood, and perhaps other materials. While you may have had to do this type of construction yourself, out of necessity, in graduate school, you will find this practice discouraged in industry for reasons of efficiency and safety, among others.

Eventually you may have a product or process ready for evaluation on equipment similar to that used in actual production. Here you should turn to the *technical service and sales service laboratories*, which are in close touch with the customers and the equipment they will be using.

If your work is related more to the applied aspects of R&D, you will probably find that the *manufacturing areas* offer much that is relevant to your research. In development work, you may be carrying out experiments on production equipment. This is expensive, and the consequences of a mistake could be disastrous. In applied research, you will be working toward the goal of transferring your technology to the development groups for implementation in the plant. In either case, an awareness of the nature of the existing manufacturing processes will allow you to design any modifications for maximum compatibility with existing plant equipment and processes.

You must maintain an awareness of the hazards and toxicity of any materials you are working with. Start with a study of the pertinent literature (59–62), and continue as needed by consulting the *safety department* and its associated laboratories. We cannot overemphasize the importance of maintaining safety in everything you do, at home as well as in the laboratory. It is essential that the safety groups evaluate the toxicity, safety, and environmental and health considerations of new chemicals, products, and processes as early as possible.

The safety department will also assist you in disposing of toxic or flammable waste material. These must *never* be thrown out or down the drain casually!

If your research leads into problems that are beyond the scope of your organization to handle, it may be more efficient to have the work done under contract with *independent research laboratories* associated with a university, foundation, or government agency. Usually management will decide when this should be done, but you should be aware of the possibility.

The effectiveness of your research will depend a great deal on how well you communicate with your colleagues and the internal and external service facilities we are describing. While they are there to inform you, it is your responsibility to advise them of what you need to know. The remaining areas in Fig. 28 deal in one way or another with communications.

The *marketing* arm of your company provides you with the means of communicating with the customers who will buy and use your products. It is a good idea, where possible, for you to visit customers' plants with a marketing representative, to find out how they do things and how your products and processes are used. Many industrial research professionals are quite reluctant to do this, though we find it hard to understand why.

Communications with the *patent and legal departments,* and its advantages to you, are discussed in Chap. 10. The value to you of the *library* and *information services* has already been mentioned. Last but not least on our list are *in-house experts* and *consultants* retained by the company. Even if these people can't solve your problem, experience has shown that just talking it over helps to make you more receptive to new ideas. Such discussions can effectively stimulate creativity. And the greater experience of in-house experts and consultants may provide immediate solutions to some of the difficulties you will encounter.

APPROACH THE PROBLEM PROPERLY

The title of this section sounds obvious, but what is the proper way to approach a first-assignment R&D problem? It depends a great deal on you and your personality traits in the areas of adaptability, receptiveness, motivation, problem-solving ability, and creativity. Let us give you an example.

When he was being interviewed for a job, one of us was asked: "What is the first thing you would do if you were given a new research assignment in an unfamiliar area?" His answer, "I would go to the library and gain some acquaintance with the subject" was met with, "You mean you wouldn't go into the laboratory to run even a single experiment to get your feet wet?" and "How could I run an experiment intelligently without first becoming familiar with some of the more important principles?"

The interviewer was clearly looking for a response indicating a man of action, who would waste no time in getting going on his own. The candidate could see little sense in this until he knew where he was going. There is some merit in each point of view. You can become too well informed prior to starting experimental work. You may overlook simple but significant factors that are not obvious, or develop a bias from the experience of previous workers. On the other hand, starting right off may lead you to waste time with meaningless experiments. Here are some guidelines that tend to follow a middle point of view:

- Develop some background knowledge in your new assignment by reviewing the literature and previous company reports. Try to find out why earlier work was successful or unsuccessful, what variables were important, and whether any significant factors were overlooked.

- Get your feet wet in the laboratory with a few simple experiments, perhaps repeating previous work. At this stage run the experiments yourself, or at least follow them very closely.
- Before you formulate a plan of attack for the next stage, talk to others who were previously involved in the same or a related project.
- Devise an initial program, taking into account all resources that can expedite your progress.
- Put your plan into action. Anticipate the results as much as possible so as to plan ahead for succeeding stages.
- Revise your program as many times as you need, based on the results as they come in.
- Be receptive and open-minded; don't discount the importance of an observation or variable until you can be sure what it means.
- Utilize your support resources properly and fully.
- Provide for and pursue alternative courses of action whenever possible; don't put all your eggs in one basket.
- Don't slack off if you are given the freedom to pursue your project with minimal direction; carry out the work conscientiously.
- Communicate and work with others.
- Keep full and accurate accounts of your research in a proper notebook, making entries immediately. Chapter 10 tells you some of the reasons why.
- Keep your supervisors informed of your progress: Don't wait for them to ask you, but tell them via brief written memos.

TRY THE EDISONIAN APPROACH

As we said earlier in this chapter, the Edisonian approach is a trial-and-error method of problem solving based on the observation of cause-and-effect relations without the development of an understanding of the underlying science. While this may seem a nonscientific way to go about things, it is certainly true that many industrial products are made successfully without a full understanding of the science behind them. It has often been said that you have to be Thomas Edison to make the Edisonian approach work. Obviously not, but it does take a receptive and creative person to employ this approach effectively.

We recommend this approach, for example, to determine the importance of process variables and the range of acceptable conditions required for operating a new process. Since most processes operate with a large number of variables, it is unlikely that these can be optimized through an Edisonian approach, but it is often possible to get off the ground and make the product in a reasonable way.

PLAN AND FOLLOW THROUGH

In some respects, industrial R&D resembles the old army game of "hurry up and wait." To run a large-scale experiment, you probably have to order raw materials and equipment ahead of time, then wait for an opportunity to fit it into the schedules of several support groups. After it has been run, you may have to wait weeks or months before all of the analyses are complete and the full results in. Even before this, you will probably be planning and scheduling the next experiments. Furthermore, you will be expected to have several projects going at all times, so that your time can be used effectively even though a delay is experienced in one of them.

To cope with this situation requires good organization and planning and effective follow-through on your part. You must learn to anticipate, plan ahead, and make full use of data as they come in. For some, it is not easy to adopt this style of research, which may be quite different from that in graduate school.

Anticipating, planning, and setting the wheels in motion is not difficult, but following through may be. It isn't easy to relate the results received in the mail today to the experiment run three months ago, especially if you have been fully occupied with other projects in the interim. Remembering is an important part of following through and project evaluation. You may find it wise to establish a "tickler" system similar to one described recently (63) to help in these stages.

BE RECEPTIVE

It sounds deceptively simple to say that you must be receptive to be successful in R&D. By this we mean you must be open-minded and observant of any factors affecting your experimental results. Perhaps those analytical data seem out of line: Was it poor sample preparation, an error in the analysis, or some significant effect in the experiment? The receptive researcher finds out. Receptivity is closely related to innovation since it is the receptive mind which recognizes the possibility of change. Our favorite account of receptivity in the innovative process, describing the development of accurate techniques for gunfire on naval vessels (64), makes fascinating reading.

DON'T IGNORE ACCIDENTS AND NEGATIVE RESULTS

Closely related to receptivity is the ability to determine the significance of the unexpected result, whether accidental or not (Fig. 29). We include

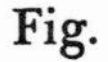

"I realize that many breakthroughs in science have come from mistakes, Halley, but nothing at all has come from the mistake I made in hiring you."

FIGURE 29. From *Ind. Res.* **16**(12), 42 (December 1974). Reprinted by permission of *Industrial Research* magazine.

negative results here because of the very great importance of finding out what cannot be done.

The establishment of technical infeasibility is often as important as the reverse. Unlike the alchemists, we think we know enough now about the fundamental characteristics of matter to abandon the search for a way to convert lead into gold. The sooner you realize that a particular project goal cannot be reached with available materials or processes, the sooner you can make the decision to terminate this attempt, and the more time and money saved from the pursuit of red herrings. Negative results from properly conceived, executed, and analyzed experiments should tell you that this is not the way to proceed and lead you to try other routes to a solution to the problem.

Scientific accidents have unquestionably led to significant changes in our society, playing important roles at the beginning of the development of electricity, telegraphy, modern photography, and antibiotics, to mention only a few (65). The researchers observing these accidents all had one thing in common: They were receptive, and traced the effect back to its cause. A recent account of the accidental discovery of a new product illustrates our point so well that we quote it verbatim in the words of the inventor, Roy J. Plunkett (66):

The Flash of Genius: Discovery of "Teflon" Polytetrafluoroethylene*

During the summer of 1938, while I was a research chemist at the Jackson Laboratory of E. I. du Pont de Nemours and Company, Penns Grove, New Jersey, I was carrying out research studies in the preparation of fluorochlorohydrocarbons.

For the solution of one of my problems, I was interested in a supply of tetrafluoroethylene. Up until this time tetrafluoroethylene had been made only in very small quantities in laboratory studies. I desired to have up to 100 pounds of this material. After carrying out some laboratory experiments, I devised a pilot plant process for producing the desired quantities of tetrafluoroethylene from dichlorotetrafluoroethane. The tetrafluoroethylene was placed in cylinders and stored in a cold storage box cooled with solid carbon dioxide.

My further research involved the reaction of tetrafluoroethylene with other chemicals to produce novel compounds.

One day with the aid of my helper, Jack Rebok, I was vaporizing tetrafluoroethylene from a small cylinder which had contained approximately 2 pounds of tetrofluoroethylene. The gaseous tetrafluoroethylene which emerged from a cylinder located on a platform scale was passed through flow meters and then led into the reacting chambers where the tetrafluoroethylene was to be reacted with other chemicals.

On this particular day, soon after the experiment started my helper called to my attention that the flow of tetrafluoroethylene had stopped. I checked the weight of the cylinder and found that it still contained a sizable quantity of material which I though to be tetrafluoroethylene. I opened the valve completely and ran a wire through the valve opening but no gas escaped. When I shook the cylinder and found there was some solid material inside, I then removed the valve and was able to pour the white powder from the cylinder. Finally, with the aid of a hack saw, the cylinder was opened and a considerably greater quantity of white powder was obtained.

It was obvious immediately to me that the tetrafluoroethylene had polymerized and the white powder was a polymer of tetrafluoroethylene.

Following this discovery, I immediately took steps to characterize the white powder and to determine ways and means by which it could be formed.

* From A. B. Garret, "Flash of Genius, 2 Teflon: Roy J. Plunkett," *J. Chem. Educ.* **39,** 288 (1962).

We feel compelled to add that the awareness of industrial safety has sharpened considerably in the intervening third of a century, and while we hope that a discovery such as Plunkett's would not be inhibited today, it would have to be made with full consideration of the safety of all concerned.

KEEP TRACK OF THE DOLLARS

All R&D projects have budgets, and you should adhere to yours as closely as possible. Revise them periodically if it becomes clear that they will be significantly under- or overspent. The periodic cost summaries prepared by the accounting department giving year-to-date expenditures and breakdowns by categories should be checked carefully, not only as a means of following the costs, but also to ensure that any errors are corrected. Estimates of project costs are usually prepared by the researcher and refined at higher levels.

Most companies place reasonable limits on the amount that a technical employee can spend on his own initiative. For example, a researcher may be able to write a requisition for a purchase costing up to $250; a supervisor, up to $500; and a laboratory head, up to $1000.

Another expense which must be controlled is that for overtime work. In most cases overtime must be approved by someone at a high level, such as the laboratory head. He will usually do so only when it is absolutely essential, and will see that the overtime is equitably distributed among the wageroll employees. As a salaried professional, you of course do not get extra pay for overtime work except in special circumstances.

TRY A BOOTLEG PROJECT

Many companies encourage their professional researchers to spend a small part of their time, say up to 10%, on a project of their own origin and interest (67). These "bootleg" or "underground" projects, carried out on the side and with minimal cost, can prove to be very valuable to the company. A recent example (68) is a new oxygen treatment for waste water which originated in a bootleg project of Jack McWhirter of Union Carbide and is beginning to have ecological repercussions throughout the world.

KEEP YOUR EDUCATION GOING

We stressed in Chap. 4 the urgency of combating technical obsolescence by taking every opportunity to continue and update your education. Reread "Sharpening Your Skills," pay special attention to the checklist on pages 78–79, and apply these ideas during your first R&D assignment.

CHALLENGES AHEAD

In this final section, we turn from the problems you face as an individual entering the industrial R&D organization to those you share, in a sense, with your management. The effective management of R&D is indeed complex: "The leader[ship] of an R&D division, it would appear, requires a veritable paragon of a human being, an ideal combination of a scientist and administrator, yogi and commissar, thinker and doer" (69). The problems this ideal leader must face include interrelated contributions from the technical, administrative, production, and marketing fields. He must face these questions, and share them with you who form his professional staff:

- How can we establish a creative and innovative atmosphere within the confines of an organized R&D division?
- How can we effectively expand our receptivity to change?
- How can we effectively transfer technology from R&D into production?
- How can we evaluate the returns from R&D?
- How do we select the right projects, and how do we know when to terminate a project?
- How do we prevent our R&D division from becoming cut off from the rest of the company and unaware of corporate goals (since boards of directors seldom include an R&D expert)?
- How do we measure the impact of R&D efforts on company progress, especially in a large, mature company?

People are the key to effective R&D, just as the company's employees overall are the key to its success. Good R&D managers spend a significant part of their time on concerns related to providing the right kind of atmosphere and organization for their professional people. Once the proper creative atmosphere has been established, it must be maintained by constant vigilance through the years of business ups and downs.

We don't intend to get into a discussion of research management, but we feel you should be aware of some of the more important problem areas in industrial R&D. We begin with creativity, since perhaps more than anything else it is a prerequisite for successful R&D.

CREATIVITY

The creative process has been defined (70) as "the emergence in action of a novel relational product, growing out of the uniqueness of the individual on the one hand, and the materials, events, people, or circumstances of his life on the other." Creativity can be fostered by establishing the proper conditions of psychological safety and freedom.

For psychological safety, one must provide acceptance of the unconditional worth of the individual, regardless of his present condition or behavior; an atmosphere free from external evaluation; and empathetical understanding. Psychological freedom requires that the individual be given the opportunity to be free and open-minded in his thinking and actions.

These views (70) are supported by the results of a recent survey on the stimulation of creativity and productivity (71) which found the most important environmental factors for stimulating creativity to be, in order of importance, recognition and appreciation, freedom to work in areas of greatest interest, broad contacts with stimulating colleagues, encouragement to take risks, opportunity to work alone rather than on a team, monetary rewards, criticism by supervisors or associates, creativity training programs, and regular performance appraisals.

It is interesting that creativity training programs and regular performance reviews are at the bottom of the list.

The same study (71) listed the most characteristic traits of a very creative scientist as enthusiasm about his work, ability to analyze and simplify a problem, drive to achieve, curiosity giving an urge to solve problems, ability to transfer concepts from one field to another, independence in thought and work, ability to stimulate his associates, broad interests, and dissatisfaction with the *status quo*.

Look again at the list of qualifications desired in job candidates by prospective employers, at the beginning of Chap. 3. Is it any wonder that the the two lists are very similar?

Providing the proper environment to foster creativity takes both time and effort of management and the proper commitment of everyone in the organization. Such an environment benefits everyone, however, regardless of his level of creativity.

It is important for industry to recognize highly creative people early, and to see that they are given assignments that will motivate them and inspire their creative genius. Ineffective management and dull, routine assignments can stifle creativity. There has developed a considerable literature on creativity and motivation in industry; we can especially recommend articles (72–78) dealing with motivation in industrial R&D, books (79–82) on motivation in a more general sense, recent articles (69, 71) exploring creativity in industry, and two excellent books (83, 84) on managing creative R&D personnel.

INNOVATION

Creativity is a key ingredient in successful innovation, "the process of carrying an idea—perhaps an old, well known idea—through the laboratory, development, production and then on to successful marketing of the product." The article (85) providing this definition of innovation notes that profitable marketing is a key result of industrial innovation. Another definition (86) makes the same point: Technical innovation is a "complex activity which proceeds from the conception of a new idea to a solution of the problem, and then to the actual utilization of a new item of economic or social value." Thus the term innovation has a special meaning beyond just something new, and should not be confused with scientific discovery or invention. Both of these are likely to be important, however, at some stages of the innovative process. Some of the characteristics of this process are listed in Table 19.

A recent study conducted by Battelle for the NSF (39, 40, summarized in 87) analyzed ten major innovations for the factors contributing most to their success. These are listed in Table 20, with the percentage of decisive events in the innovative process for which that factor was judged

TABLE 19

Characteristics of the Innovative Process*

Early Recognition of Need—Recognition of the need for the innovation generally occurs prior to the availability of the technological means for satisfying the need.

Independent Inventor—The independent inventor, working on his own behalf, is often important in the initiation of the process.

Technical Entrepreneur—the technical entrepreneur is often important to the successful culmination of the innovation.

External Invention—Many innovations arise from inventions which originate outside the organization that developed the innovation.

Government Financing—Government financing is important in many innovations.

Informal Transfer of Knowledge—Innovations are facilitated by informal transfer of knowledge, much more than through formal channels of communication.

Supporting Invention—Innovations generally require additional inventions beyond the initiating invention.

Unplanned Confluence of Technology—The innovative process is frequently facilitated by an unplanned confluence of technology.

* From (39, 40).

TABLE 20

Important Factors Affecting Innovation*

Factors	Percentage of decisive events
Recognition of technical opportunity	87
Recognition of the need	69
Internal R&D management	66
Management venture decision	62
Availability of funding	62
Technical entrepreneur	56
In-house colleagues	51
Prior demonstration of feasibility	49
Patent/license considerations	47
Recognition of scientific opportunity	43
Technology confluence	36
Technological gatekeeper	30
Technology interest group	29
Competitive pressures	25
External direction to R&D personnel	16
General economic factors	16
Health and environmental factors	15
Serendipity	12
Formal market analysis	7
Political factors	5
Social factors	4

* From (39, 40).

important. (In retrospect, one might argue that formal market analysis was bound to be rated low, because such an analysis usually is done only once, and does not continue through the innovative period. But the same argument might apply to management venture decision, which ranks high.) From Tables 19 and 20, it is apparent that R&D management plays a very important role in innovation. Management policies can either foster or stifle this process.

Innovators are a special breed of people. They "usually have a broad range of interests and experiences, they tend to be non-conformists, are self-starters, don't give up after initial failures, perform a relatively large volume of work, are sensitive to their environment, may or may not react favorably under pressure, and are generally dissatisfied with the status quo" (85).

Surprisingly, the record of large companies for innovation is not too good. Out of 88 major innovations in this century, only 23 came from large corporations (86). One might have expected a better record in view of the extensive R&D programs of large companies. It is not clear why individual inventors, small companies, independent laboratories, and researchers in academic and government laboratories were more innovative. Perhaps it is because innovation is, by its very nature, associated with change. Most large companies, as we have noted before, seem to be almost resistant to change and new ideas, even though they may themselves have developed out of earlier innovations. There is an inertia, perhaps motivated by fear, which must be overcome before new ideas can be accepted which seem to threaten the firm's stable existence. Change in large corporations just doesn't take place over night, whereas small corporations are more flexible and receptive to new ideas (88).

More insight into innovation in industry can be cultivated by reading refs. 39, 40, and 85–95.

PROJECT SELECTION

One of the thorniest problems faced by industrial R&D management is the selection of the projects to be worked on. As David J. Rose of MIT puts it (96),

> Good science and technology are arts of the rarest kind, though often given over to mediocre practitioners, and sometimes prostituted for daily pay. The art and the success come mainly in choosing which path to follow among many possible ones: far more technical and scientific directions always exist than can be followed. As in all things, initial choices must be made largely in intuition, before the facts or the consequences are fully known or understood; therein lies the art.

Art may be the key to successful project selection, but a tremendous amount of effort has been expended, especially in recent years of rising costs, to predict scientifically the selection of projects with a high chance of success. The problem lies in evaluating the multiplicity of criteria and influences on which the chances of success depend. Among these are (97) corporate guidelines and attitudes (business objectives, desire for innovation, extent of long-range commitment, courage in risk taking), corporate capabilities (marketing capabilities, position relative to competitors, state of technology required), R&D policies (willingness to accept technical risk, optimum project size, utilization of outside support), and economic factors (R&D costs, commercialization costs, commercial value, profitability). Among these criteria for project success, most attention is usually paid to the economic justification, since economic performance is at the heart of any business enterprise.

In the evaluation of these criteria, certain guidelines are important (97): Since a limited resource (R&D money) is to be allocated among a presumably unlimited group of projects, with only a few of the projects initiated likely to reach commercialization, it is better to risk rejecting a good project rather than to risk undertaking a poor one. Value and cost should, however, be evaluated independent of risk, since the latter can probably be controlled during the ensuing R&D if the value is high enough. Commercialization costs must be considered carefully, since they will likely outweigh R&D costs eventually. The time scale of the project must also be considered, and the commercial goals related to needs at the time of their fulfillment, not of the start of the project. Finally, the degree to which it meets the needs of the company is the key to final justification of an R&D project.

Some companies formalize the evaluation of new proposals with the aid of a checklist, like the one shown in Table 21. Although such a list is useful, especially as an aid to keeping evaluations consistent from one project to the next and seeing that major points are not overlooked, it is unlikely that a sophisticated company would rely on this approach alone. It can be very useful, however, to the researcher who develops an idea which he thinks might lead to a project proposal.

Another technique for evaluating the economic factors associated with a project is to use one of many formulas developed to weigh these quantitatively (9, 10). One calculates an index of value or success, which may be simple or complicated, by a formula like those shown in Fig. 30.

The extent to which detailed consideration of a project is required, whether the checklists or formulas just described are used or not, depends on the size of the project (98). In exploratory research, where the initial effort is small, the researcher may prepare a short informal writeup of a project for approval by his immediate supervisor. A project that will require the effort of several men for 1–2 years obviously requires more detailed consideration and approval at a much higher level, but still may involve only R&D. When the commitment of major corporate effort is in question, the selection procedure becomes much more rigorous, with full marketing and economic evaluations required. At this stage it may take a team just to pull together all the relevant data for the decision.

Further information on project selection is given in refs. 9, 10, and 97–100.

TABLE 21

A Checklist for Evaluating R&D Projects*

	Very favorable	Favorable	Average	Unfavorable	Very Unfavorable
Financial					
Estimated annual sales of new product	___	___	___	___	___
Time to reach estimated sales volume	___	___	___	___	___
Ratio of annual sales: R&D costs	___	___	___	___	___
Ratio of total costs: annual savings	___	___	___	___	___
Return on sales	___	___	___	___	___
Return on fixed capital	___	___	___	___	___
Return on total investment	___	___	___	___	___
R&D investment payout time	___	___	___	___	___
Fixed capital investment payout time	___	___	___	___	___
Profit in first year of production	___	___	___	___	___
Research and Development					
Chance of technical success	___	___	___	___	___
Technical novelty	___	___	___	___	___
Potential know-how gain	___	___	___	___	___
Relation to company's present know-how	___	___	___	___	___
Time to develop product	___	___	___	___	___
Manpower needed	___	___	___	___	___
Lab and pilot plant equipment needed	___	___	___	___	___
Competitive technical activity	___	___	___	___	___
Patent status	___	___	___	___	___
Production					
Process advantage	___	___	___	___	___
Process versatility	___	___	___	___	___
Process familiarity	___	___	___	___	___
Compatibility with present operations	___	___	___	___	___
Equipment availability	___	___	___	___	___
Raw material availability	___	___	___	___	___
By-product outlets	___	___	___	___	___
Waste disposal	___	___	___	___	___
Corrosion potential	___	___	___	___	___
Hazard potential	___	___	___	___	___
Freight position	___	___	___	___	___

	Very favorable	Favorable	Average	Unfavorable	Very Unfavorable
Marketing					
Product advantage	___	___	___	___	___
Product competition	___	___	___	___	___
Market size	___	___	___	___	___
Market stability	___	___	___	___	___
Market permanence	___	___	___	___	___
Cyclical and seasonal demand	___	___	___	___	___
Number of potential customers	___	___	___	___	___
Market growth rate	___	___	___	___	___
Company known in potential markets	___	___	___	___	___
Compatibility with present products	___	___	___	___	___
Suitable marketing organization available	___	___	___	___	___
Market development requirements	___	___	___	___	___
Promotional requirements	___	___	___	___	___
Technical service requirements	___	___	___	___	___
Time required to become established in market	___	___	___	___	___
Product variations and modifications required	___	___	___	___	___
Difficulty of copying or substituting product	___	___	___	___	___
Export potential	___	___	___	___	___
Possibility of a captive market	___	___	___	___	___
Licensing potential	___	___	___	___	___
Corporate Position					
Relation to company objectives	___	___	___	___	___
Required corporate size	___	___	___	___	___
Advertising or prestige value	___	___	___	___	___
Effect on purchasing other materials	___	___	___	___	___
Effect on present customers	___	___	___	___	___
Operating departments' desire or enthusiasm	___	___	___	___	___
Other Factors					
	___	___	___	___	___
	___	___	___	___	___
	___	___	___	___	___
	___	___	___	___	___
	___	___	___	___	___

* From (10). Reprinted by permission of *Chemical & Engineering News.*

Simple Index of Return

$$\text{Index} = \frac{\text{estimated value of research if successful} \times \text{estimated chance of technical success}}{\text{estimated cost of the research project}}$$

Product Life Index of Return

$$\text{Index} = \frac{\text{\% chance of technical success} \times \text{\% chance of commercial success} \times \text{estimated unit sales per year} \times \text{profit per unit} \times \text{life of product in years}}{\text{estimated cost of the project}}$$

Profitability Index

$$\text{Index} = \frac{\text{\% chance of technical success} \times \text{\% chance of commercial success} \times \text{annual sales volume in units} \times \text{unit sales price} \times \text{static market life}}{\text{R \& D cost} + \text{production-engineering developing cost} + \text{marketing development cost}}$$

Index of Relative Worth

$$\text{Index} = \frac{\text{\% chance of technical success} \times \text{\% chance of commercial success---nature of project} \times \text{\% chance of commercial success---economic} \times \text{net earnings over life of product} \times \text{present worth factor} \times \text{required new capital investment}}{\text{estimated cost of research, development, and engineering}}$$

Calculated-Risk Index

$$\text{Index} = \frac{\text{sum of all annual net incomes to be earned} \times \text{risk factor}}{\text{investment in product development, capital assets, and market development}}$$

FIGURE 30. Some formulas for evaluating the economics of new R&D projects (9).

PROJECT TERMINATION

Deciding when to stop work on a project is, if anything, more difficult than deciding what to work on. The dilemma here is whether to continue to pour money into a project with no chance of success, or to cut it off too soon, when just a little more effort would have turned the corner. There is no clear answer to this problem; perhaps the best safeguard is critical periodic review and a harsh attitude. If nothing is done, the project will continue to run; it is more difficult to call a halt. Perhaps the only time when stopping a project is easy is when the researchers themselves admit that they have run out of ideas or enthusiasm.

The crystal ball can sometimes be very, very cloudy with respect to project termination. Ironically, in 1964 (10) Crawford H. Greenewalt, then chairman of Du Pont's Board of Directors, used Corfam as an example of when to terminate a project. At that time Corfam was at the height of its success, and Mr. Greenewalt was referring to three or four previous attempts to develop a leather substitute within the company. "But we were able to recognize that the ideas were not good enough before we expended vast sums of money," he said at that time. "After each failure we simply put the problem away until new technology, new material, or new ideas became available. Had we conducted research unremittingly for the entire 30-year period, I doubt that we would have been successful any earlier, and we would have spent so many research dollars that even a brilliant commercial success would not have bailed us out."

Alas, as we saw on pages 140–141, the full story on Corfam was not in at that time. Who is to say when the decision to stop *that* project should have been made?

REFERENCES

1. C. E. Kenneth Mees and John A. Leermakers, *The Organization of Industrial Scientific Research,* 2nd ed., McGraw-Hill, New York, 1950.
2. National Science Foundation, *Research and Development in Industry, 1967,* Report NSF 69–28, July 1969, Washington, D.C.*
3. Alfred E. Brown, "New Definitions for Industrial R&D," *Res. Manage.* **15** (5), 55–57 (Sept. 1972).
4. Harvey Brooks, "Applied Research Definitions, Concepts, Themes," pp. 21–55 in National Academy of Sciences, *Applied Science and Technological Progress,* report

* Available from the Superintendent of Documents, U.S. Government Printing Office, Washington, D.C. 20402.

to the Committee on Science and Astronautics, U. S. House of Representatives, Washington, D.C., June 1967.*

5. Anon., "Employment Up as Industrial R&D Funding Improves," *Res./Dev.* **25** (3), 10 (1974).
6. William F. Fallwell, "Chemical Company R&D Spending Stays Strong," *Chem. & Eng. News* **53** (2), 8–9 (Jan. 13, 1975).
7. William Lerner, ed., *Statistical Abstract of the United States, 1973,* U.S. Dept. of Commerce, Washington, D.C., 1973.*
8. National Science Foundation, *Science Indicators 1972,* Stock No. 3800-00146, U.S. Government Printing Office, Washington, D.C., 1973.*
9. Robert E. Seiler, *Improving the Effectiveness of Research and Development,* McGraw-Hill, New York, 1965.
10. David M. Kiefer, "Winds of Change in Industrial Chemical Research," *Chem. & Eng. News* **42** (12), 88–109 (March 23, 1964).
11. Edwin M. Lindsay, "Financial Management of R&D: Planning and Budgeting, Project Authorization, Financial Reporting," *Res. Manage.* **14** (4), 58–66 (July 1971).
12. William E. Souder, "Budgeting for R&D," *Bus. Horiz.* **13** (3), 31–38 (June 1970).
13. William C. Goggin, "How the Multidimensional Structure Works at Dow Corning," *Harvard Bus. Rev.* **52** (1), 54–65 (Jan.–Feb. 1964).
14. Ronald D. Clark, "Can the Chemical Industry Afford Research?" *Chemtech* **2**, 656–659 (1972).
15. Charles B. McCoy, "A View from Industry—Chemistry in the Service of Mankind," speech at Purdue University, April 27, 1973. E. I. du Pont de Nemours and Co., Wilmington, Del., 1973.
16. Roger W. Gunder, "View from the Top," *Chemtech* **1**, 130–131 (1971).
17. Robert A. Charpie, "What Management Expects from R&D Today," *Res. Manage.* **16** (2), 7–9, (March 1973).
18. Robert S. Ingersoll, "What the Corporate Executive Expects of R&D," *Res. Manage.* **15** (2), 52–58 (March 1972).
19. John Lobb, "What Chief Executives Expect from the R&D Department," *Res. Manage.* **14** (3), 74–79 (May 1971).
20. Donald W. Collier, "How Can Industrial Research Meet the Changing Expectations of Corporate Management?" *Res. Manage.* **14** (3), 56–63 (May 1971).
21. E. C. Galloway "Evaluating R&D Performance—Keep It Simple," *Res. Manage.* **14** (2), 50–58 (March 1971).
22. Robert E. Gee, "The Opportunity Criterion—A New Approach to the Evaluation of R&D," *Res. Manage.* **15** (3), 64–71 (May 1972).
23. Shirley A. Edwards and Michael W. McCarrey, "Measuring the Performance of Researchers," *Res. Manage.* **16** (1), 34–41 (Jan 1973).
24. James K. Brown and G. Clark Thompson, "Company R&D: Status and Outlook," *Conf. Board Rec.* **3** (2), 7–17 (1966).

* Available from the Superintendent of Documents, U.S. Government Printing Office, Washington, D.C. 20402.

25. Harold Stieglitz and Allen R. Janger, "Research and Development," Chap. 11 in *Top Management Organization in Divisionalized Companies,* Personnel Policy Study No 195, National Industrial Conference Board, New York, 1965.
26. R. E. Marshak, "Basic Research in the University and Industrial Laboratory," *Science* **154**, 1521–1524 (1966).
27. Arthur M. Bueche, "Industrial Basic Research," *Chem. & Eng. News* **47** (35), 57–59 (Aug. 25, 1969).
28. James Tait Elder, "Basic Research in Industry: Appraisal and Forecast," *Res. Manage.* **6** (1), 5–14 (Jan. 1963).
29. R. S. Morrison, "An Intellectual Endeavor," *Chem. & Eng. News* **47** (35), 52–56 (Aug. 25, 1969).
30. William F. Gresham, "What I Have Learned—About Exploratory Research," *Res. Manage.* **17** (2), 8–10 (March 1974).
31. Glenn L. Bryan, "The Role of Basic Research in the Total R&D Process," *Res. Manage.* **16** (1), 29–33 (Jan. 1973).
32. Thomas J. Hogan, "The Present and Future of Company-Funded Basic Research," *Res. Manage.* **16** (3), 26–29 (May 1973).
33. David Allison, ed., *The R&D Game,* MIT Press, Cambridge, 1969.
34. James Brian Quinn, *Yardsticks for Industrial Research,* Ronald, New York, 1959.
35. David B. Hertz, ed., *Research Operations in Industry,* King's Crown, New York, 1953.
36. Eugene H. Kone and Helene J. Jordan, eds., *The Greatest Adventure: Basic Research That Shapes Our Lives,* Harper & Row, New York, 1974.
37. Donald N. Frey and J. E. Goldman, "Applied Science and Manufacturing Technology," pp. 255–295 in ref. 4; C. Guy Suits and Arthur M. Bueche, "Cases of Research and Development in a Diversified Company," pp. 297–346 in ref. 4.
38. John Bardeen, "Research Leading to the Point-Contact Transistor," *Science* **126,** 105–112 (1957).
39. Battelle Columbus Laboratories for the National Science Foundation, *Science, Technology, and Innovation,* Report NSF-BCL-C-667-73-ABR; PB228-509, National Technical Information Service, U.S. Dept. of Commerce, Springfield, Va., 1973.
40. Battelle Columbus Laboratories for the National Science Foundation, *Interactions of Science and Technology in the Innovation Process: Some Case Studies,* Report NSF-BCL-C-667-73-COMP; PB228-508, National Technical Information Service, U.S. Dept of Commerce, Springfield, Va., 1973.
41. John E. Bujake, Jr., "Ten Myths about New Product Development," *Res. Manage.* **15** (1), 33–42 (Jan. 1972).
42. William T. Constandse, "How to Launch New Products," *MSU Bus. Top.,* 29–34 (Winter 1971).
43. Albert V. Bruno, "New Product Decision Making in High Technology Firms," *Res. Manage.* **16** (5), 28–31 (Sept. 1973).

* Available from the Superintendent of Documents, U.S. Government Printing Office, Washington, D.C. 20402.

44. Anon., *Getting the Most from Product Research and Development,* Special Report No. 6, American Management Association, New York, 1956.

45. Anon., *Research, Management, and New Product Development,* Arthur D. Little, Cambridge (no date).

46. Morgan B. MacDonald, Jr., *Appraising the Market for New Industrial Products,* Business Policy Study No. 123, The Conference Board, New York, 1967.

47. Victor J. Danilov, ed., *New Products—And Profits,* Industrial Research, Beverly Shores, Ind., 1969.

48. Harry B. Watson, *New-Product Planning,* Prentice-Hall, Englewood Cliffs, N.J., 1969.

49. Philip Gisser, *Launching the New Product,* American Management Association, New York, 1972.

50. Anon., *The D of Research and Development (This is Du Pont 30),* E. I. du Pont de Nemours and Co., Wilmington, Del., 1966.

51. Fred W. Billmeyer, Jr., *Textbook of Polymer Science,* 2nd ed., Wiley-Interscience, New York, 1971.

52. Guy Alexander, *Silica and Me—The Career of an Industrial Chemist,* American Chemical Society, Washington, D.C., 1973.

53. Gilbert Burck, "Du Pont 'Gave Away' Billions—and Prospered," *Fortune* **87** (1), 68–75 (Jan. 1973).

54. Harley H. Bixler, *The Manufacturing Research Function,* AMA Research Study No. 60, American Management Association, New York, 1963.

55. Ray Hyman and Barry Anderson, "Solving Problems," pp. 90–105 in ref. 33.

56. E. L. Yuan, "The Structure and Property Relations of Poromeric Materials," presented at the 159th National ACS meeting, Houston, Texas, Feb. 22–27, 1970; Anon., "Poromerics: How They're Manufactured," *Chem. & Eng. News* **48** (10), 62–63 (March 9, 1970).

57. Anon., "Du Pont: End of Corfam," *Chem. & Eng. News* **49** (12), 10 (March 22, 1971).

58. Anon., "Du Pont, Dow Offer Analytical Services," *Chem. & Eng. News* **50** (5), 24 (Jan. 31, 1972).

59. N. Irving Sax, ed., *Dangerous Properties of Industrial Materials,* 3rd ed., Reinhold, New York, 1968.

60. Marion N. Gleason, Robert E. Gosselin, Harold E. Hodge, and Roger P. Smith, *Clinical Toxicology of Commercial Products, Acute Poisoning,* Williams and Wilkins, Baltimore, 1969.

61. Herbert E. Christensen, ed., *The Toxic Substances List, 1973 Edition,* U.S. Dept. of Health, Education and Welfare, Rockville, Md., 1973.*

62. *Merck Index: An Encyclopedia of Chemicals and Drugs,* 8th ed., Merck, Sharpe, and Dohme, Rahway, N.J., 1968.

63. B. J. Luberoff, "The Tickler," *Chemtech* **4**, 321 (1974).

64. Elting E. Morison, *Men, Machines, and Modern Times,* MIT Press, Cambridge, 1966.

* Available from the Superintendent of Documents, U.S. Government Printing Office, Washington, D.C. 20402.

65. Jerome S. Meyer, *Great Accidents in Science That Changed the World,* Arco Publishing, New York, 1967.

66. A. B. Garrett, "The Flash of Genius, 2. Teflon: Roy J. Plunkett," *J. Chem. Educ.* **39,** 288 (1962).

67. John D. Aram, "Innovation via the R&D Underground," *Res. Manage.* **16** (6), 24–26 (Nov. 1973).

68. John F. Menahan, "The Chemical Innovators 12: Jack McWhirter, New Era for an Old Idea," *Chem. & Eng. News* **49** (17), 31–33 (April 26, 1971).

69. E. Bruce Peters, "Overcoming Organizational Constraints on Creativity and Innovation," *Res. Manage.* **17** (3), 29–33 (May 1974).

70. Carl R. Rogers, *On Becoming a Person,* Houghton Mifflin, Boston, 1961, pp. 347–359.

71. S. M. Parmerter and J. D. Garber, "Creative Scientists Rate Creativity Factors," *Res. Manage.* **14** (6), 65–70 (Nov. 1971).

72. George F. Farris, "Motivating R&D Performance in a Stable Organization," *Res. Manage.* **16** (5), 22–27 (Sept. 1973).

73. George C. Bucher and John E. Reece, "What Motivates Researchers in Times of Economic Uncertainty?" *Res. Manage.* **15** (1), 19–32 (Jan. 1972).

74. George C. Bucher and Richard C. Gray, "The Principles of Motivation and How to Apply Them," *Res. Manage.* **14** (3), 12–23 (May 1971).

75. Fred Landis, "What Makes Technical Men Happy and Productive?" *Res. Manage.* **14** (3), 24–42 (May 1971).

76. Earl R. Gomersall, "Current and Future Factors Affecting the Motivation of Scientists, Engineers, and Technicians," *Res. Manage.* **14** (3), 43–50 (May 1971).

77. John H. Dessauer, "How a Large Corporation Motivates Its Research and Development People," *Res. Manage.* **14** (3), 51–55 (May 1971).

78. Frederick Herzberg, "One More Time: How Do You Motivate Employees?" *Harvard Bus. Rev.* **46** (1), 53–62 (Jan.–Feb. 1968).

79. Abraham H. Maslow, *Motivation and Personality,* 2nd ed., Harper & Row, New York, 1970.

80. John E. Hall, *Psychology of Motivation,* Lippincott, Chicago, 1961.

81. Frederick Herzberg, Bernard Mausner, and Barbara B. Snyderman, *The Motivation to Work,* Wiley, New York, 1959.

82. Saul W. Gellerman, *Motivation and Productivity,* American Management Association, New York, 1963.

83. Eugene Raudsepp, *Managing Creative Scientists and Engineers,* Macmillan, New York, 1963.

84. C. George Evans, *Supervising R&D Personnel,* American Management Association, New York, 1969.

85. Anon., "Top Research Managers Speak Out on Innovation," *Res. Manage.* **13** (6), 435–443 (Nov. 1970).

86. Sumner Myers and Donald G. Marquis, *Successful Industrial Innovations,* National Science Foundation Report 69-71, U.S. Government Printing Office, Washington, D.C., 1969.*

87. Samuel Globe, Girard W. Levy, and Charles M. Schwartz, "Key Factors and Events in the Innovative Process," *Res. Manage.* **16** (4), 8–15 (July 1973).

88. Donald A. Schon, "The Fear of Innovation," pp. 119–134 in ref. 33.

89. Michael Michaelis and William D. Carey, *Barriers to Innovation in Industry,* Report NSF-RA-X-73-003; PB-229-898, U.S. Government Printing Office, Washington, D.C., 1973.*

90. John Jewkes, David Sawers, and Richard Stillerman, *The Sources of Invention,* 2nd ed., Norton, New York, 1970.

91. J. Langrish, M. Gibbons, W. G. Evans, and F. R. Jevons, *Wealth from Knowledge —A Study of Innovation in Industry,* Halsted-Wiley, New York, 1972.

92. John A. D. Cropp, David C. Harris, and Edward S. Stern, *Trade in Innovation: The Ins and Outs of Licensing,* Wiley-Interscience, New York, 1970.

93. James R. Bright, *Research, Development, and Technological Innovation,* Irwin, Homewood, Ill., 1964.

94. J. A. Morton, *Organizing for Innovation,* McGraw-Hill, New York, 1971.

95. Robert C. Dean, Jr., "The Temporal Mismatch—Innovation's Pace vs. Management's Time Horizon," *Res. Manage.* **17** (3), 12–15 (May 1974).

96. David J. Rose, "New Laboratories for Old," *Daedalus* **103** (3), 143–155 (Summer 1974).

97. Wilson M. Whaley and Robert A. Williams, "A Profits-Oriented Approach to Project Selection," *Res. Manage.* **14** (5), 25–37 (Sept. 1971).

98. Eric S. Whitman and Edward F. Landan, "Project Selection in the Chemical Industry," *Res. Manage.* **14** (5), 56–61 (Sept. 1971).

99. Robert E. Gee. "A Survey of Current Project Selection Practices,' *Res. Manage.* **14** (5), 38–45 (Sept. 1971).

100. Ross Clayton, "A Concept Approach to R&D Planning and Project Selection," *Res. Manage.* **14** (5), 68–74 (Sept. 1971).

* Available from the Superintendent of Documents, U.S. Government Printing Office, Washington, D.C. 20402.

CHAPTER 7

MANUFACTURING

Our objective in this chapter is to present what we consider to be the major facts that you, a prospective or new industrial scientist or engineer, should know about the manufacturing segment of the American chemical industry. Because the output of the manufacturing division is more readily visible than a company's efforts in R&D or marketing, the overall description of the chemical industry which we presented in Chap. 2 largely reflects what goes on in manufacturing. The two chapters overlap to some extent, but with differing emphasis: In Chap. 2, our facts and figures referred to the company as a whole, whereas here we single out the manufacturing arm for closer scrutiny.

Almost independent of the type of industry or product, the typical life-cycle relationships shown in Fig. 19 on p. 116 apply, and the chemical industry serves as an example in illustrating some general manufacturing concepts that apply across the board. Thus, technology originating in R&D is transferred to the manufacturing development areas for scaleup and production. These groups also play an important role in improving products and processes to maintain profitability and extend the useful life of the product. The product is then turned over to marketing, which transfers it to the consumer.

In the scheme of things, it happens that more of the job opportunities open to new employees in manufacturing are suited to engineers; thus a larger portion of the material in this chapter, in contrast to some others, is similarly directed to the engineer rather than the scientist.

MANUFACTURING IN THE CHEMICAL INDUSTRY

PRODUCTS

As we mentioned briefly in Chap. 2, the chemical industry manufactures three types of products:

- Basic chemicals, which are utilized within the chemicals and allied products industry to produce other chemical products.

- Goods, such as synthetic fibers and elastomers, which are fabricated into end products by other industries.
- Chemical end products, either unformulated or formulated compositions, such as drugs, detergents, fertilizer, or antifreeze, which are used directly by the consumer.

Some of the basic chemicals produced in largest quantities in the United States in 1973 are indicated in Table 22. We list the top dozen, plus another handful from the top 50. One might think of all these as basic chemicals, but many of them (such as oxygen and ethylene glycol) are chemical end products as well, while a surprisingly large number are monomers or intermediates in the production of monomers which, on polymerization, yield plastic, fiber, and elastomer goods. Since these high-volume chemicals represent a very substantial portion of the output of the chemical industry, it would seem important for the chemist and chemical engineer to know something about the reactions and processes for making them. In our day (at least for the elder of the authors), this was taught as part of the undergraduate curriculum, but it is dismaying to see how little of it is included today. Perhaps it is more important to learn quantum mechanics and molecular orbital theory, but we have some reservations.

TABLE 22

Some of the 50 Largest-Volume Chemicals Produced in the United States in 1973*

Rank	Chemical	Production†	Rank	Chemical	Production†
1	Sulfuric acid	63	16	Methanol	7.1
2	Oxygen	32	17	Toluene	6.8
3	Ammonia	31	19	Formaldehyde	6.2
4	Ethylene	22	20	Styrene	6.0
5	Sodium hydroxide	21	22	Vinyl chloride	5.4
6	Chlorine	21	23	Hydrochloric acid	4.8
7	Nitrogen	16	26	Butadiene	3.7
8	Sodium carbonate	15	27	Carbon black	3.5
9	Nitric acid	15	28	Ethylene glycol	3.3
10	Ammonium nitrate	14	30	Carbon dioxide	2.8
11	Phosphoric acid	13	31	Dimethyl terephthalate	2.7
12	Benzene	11	34	Cyclohexane	2.4
13	Propylene	8.8	40	Acetone	2.0
15	Urea	7.1	41	Ethanol	2.0

* Data from (1).

† Billions of pounds.

Partly because polymer science is one of our specialties and has been overlooked in most universities, we wish to point out its industrial importance by noting the large production volumes of such monomers as ethylene, propylene, styrene, vinyl chloride, and butadiene, and such precursors or intermediates as ethylene glycol and dimethyl terephthalate (polyester fibers and film) and cyclohexane (the nylons). The industries producing the resulting polymeric goods, the intermediates leading up to them, and the end products made from them, provide employment for over half of the chemists and chemical engineers working in industry in the United States today.

References 1–5 provide detailed information about the chemicals and allied products industry.

GROWTH

Since the economy appears to have gone completely berserk in the latter part of 1974, we retreat to the comparative stability of 1973 to indicate some growth trends in the production of chemicals up to that time (1):

- Plastics production increased by about 10% over 1972, continuing a trend of many years (6).
- Synthetic fiber production jumped 14%, with production running at about 93% of plant capacity.
- Paint production fell off around 2.4%.
- Drug sales posted the usual large annual increase, up 11% worldwide over 1972.
- Fertilizer consumption continued high, up 4% over 1972.
- Synthetic rubber production in 1973 was 7% above 1972.

Most of the annual growth rates in these products have, however, turned sharply down after peaking in 1972 or early 1973 (Fig. 31), and the gains following the 1970 recession were more than wiped out by the end of 1974. However, production in the chemical industry continued to remain strong relative to that in all manufacturing industries. For the decade 1963–1973, chemicals and allied products production gained at an annual rate of 8.4%, compared to 5.1% for all manufacturing.

MANPOWER

In Table 5, page 191, we showed that only 16% of all industrial chemists were directly employed in production in 1970. Adding a few percent to account for management in manufacturing, it is seen that only about one in every five employees in the chemical industry is directly con-

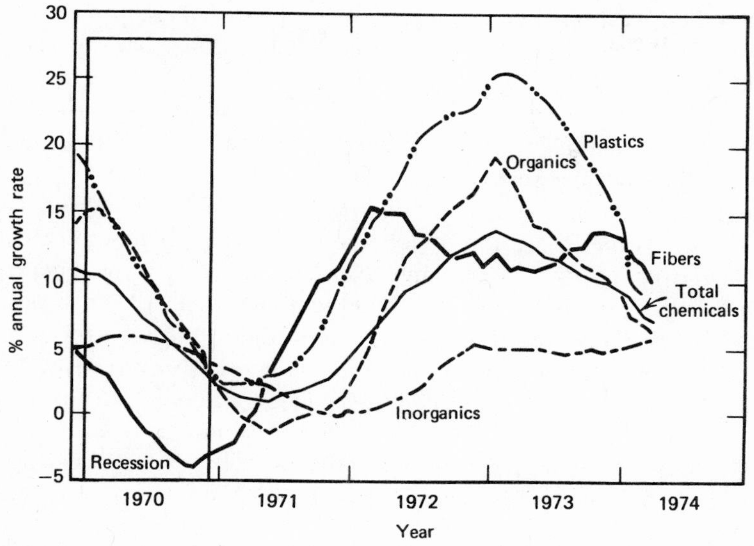

FIGURE 31. Chemical growth rates, fluctuating up and down with the economy (1,7). (Reprinted by permission of *Chemical & Engineering News.*)

cerned with manufacturing. This low figure reflects the technology-intensive nature of many chemical plants, in which large quantities of products are manufactured by a small number of workers.

PLANTS

Chemical manufacturing plants can be divided into three categories (4), depending on the nature of the operations involved. Size, labor requirements, and investment vary considerably among the three.

Large *dedicated* plants are used for the high-volume production of a single chemical. Most use continuous processes, which require very little labor to operate. Plants of this type are relatively expensive, but scaleup to larger and larger sizes leads to greater economy of investment and operation. The minimum economic size of some dedicated plants is phenomenally large, with annual capacities in the hundreds of millions of pounds per year. In such a plant, for example a synthetic rubber plant, the average annual production, in terms of sales dollars, is about $85,000 per employee.

A second type of plant is based on a specific type of chemical conversion, such as nitration, hydrogenation, diazotization, or esterification, which can be applied to any of a variety of chemicals. These *general-*

purpose synthesis (often, "pots and pans") plants contain equipment for batch unit operations involving the conversion and the purification of the reactants and products, for example filtration, distillation, and crystallization. More labor is required for these plants, but the investment cost is smaller, as is their size in general. An employee in such a plant, for example in the dye industry, generates on the average about $66,000 of sales of product per year.

Plants in which mixing, dispersion, and conversion of physical form, rather than synthesis, are the major manufacturing processes are classed as *formulating plants*. They range from small garage-type operations, through large-scale explosives and detergent plants, to plants producing molded or extruded plastics and synthetic fibers. Most of these plants require relatively high labor input and low capital investment. In the explosives industry, for example, the average annual output per employee is only some $22,000 in sales dollars of product.

Plant design and specification is usually done by the engineering staff in a large company, but they often utilize independent consulting firms to assist in the engineering work and do the construction; for smaller companies, these engineering consulting firms may do the whole job. Some of them do considerable R&D to keep abreast of their fields of specialization, and can provide high-volume plants on a turnkey basis—that is, designed, constructed, and put into operation with a guaranteed production capacity.

Capital spending for new plants skyrocketed to new highs in 1974, in a trend predicted to continue through 1975 despite the recession (Fig. 32). Because of inflation, the real increase is only about half of that depicted in the figure, and represents only a slight increase when considered as a percent of sales; traditionally, spending for new plants has been 10–12% of sales dollars. Whereas a few years ago sales of $1 billion per year were phenomenal for any company in the chemical industry, in 1975 three companies (Dow, Du Pont, and Union Carbide) each planned to spend more than that amount for new construction.

IMPORTS AND EXPORTS

Continuing a trend of many years, the United States exported more chemicals than it imported by almost 60% in 1973 (1) (petroleum as a fuel is not included in this figure). Imports will continue to play an important role in chemical production as raw materials—many of which can be supplied only by foreign sources—become more scarce.

World chemical trading is dominated by the European Economic Community; the share of the United States in world chemical exports

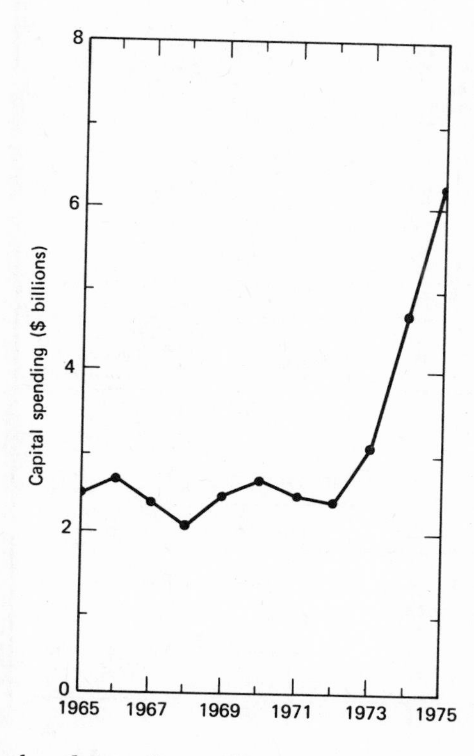

FIGURE 32. Capital spending in the chemical industry through 1974, with an estimate for 1975 (8). (Reprinted by permission of *Chemical & Engineering News*.)

has dropped steadily, from 27% a decade ago to around 18% in 1973 (1). Twelve countries of Western Europe, Japan, Canada, and the United States account for 80% of the world export of manufactured goods to foreign markets. All these countries are important customers for U.S. chemical commodities and important suppliers of raw materials.

ORGANIZATION

Recalling the role of the manufacturing division in the overall organization of a company, as indicated in Fig. 22, p. 122, let us now turn our attention to the structure of such a division. An organization chart typical of a large manufacturing division is shown in Fig. 33. Just as in an R&D division there are service groups as well as those doing actual research, so in a manufacturing division there are groups that don't manufacture anything. They serve two major purposes: to provide various production-related services, and to provide the planning and control needed to improve quality and output in their division. In Fig. 33 we have placed these groups (except for the administrative functions such as purchasing and the personnel functions, both nontechnical) into the

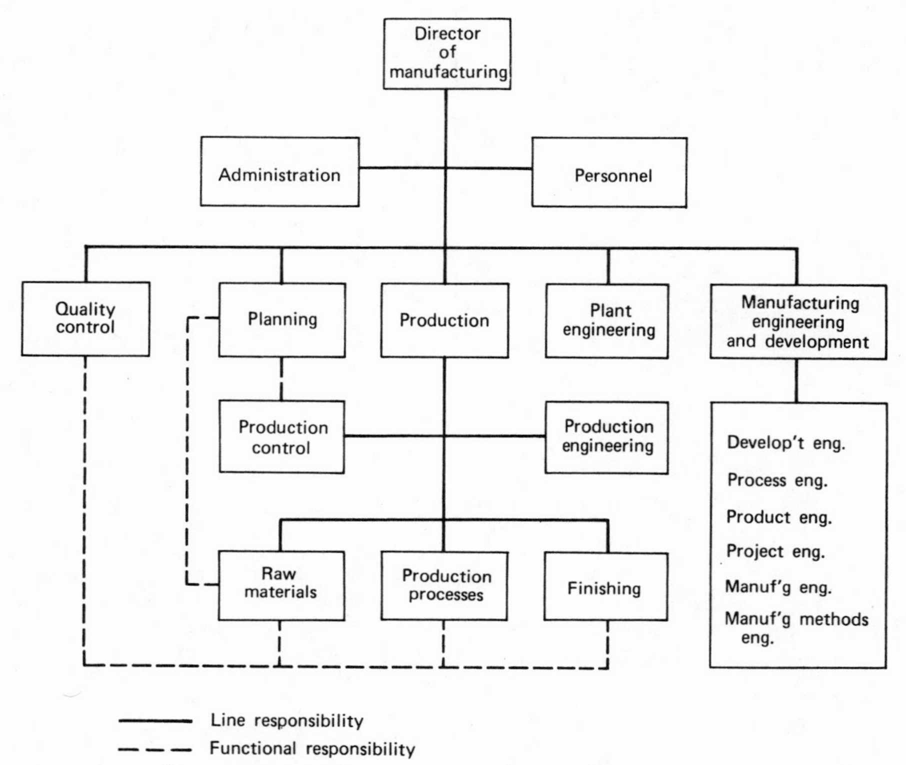

FIGURE 33. Organization of a typical manufacturing division.

four categories of planning, quality control, manufacturing engineering and development, and plant engineering. We shall examine each in detail.

These manufacturing staff activities fall into three broad areas (9). One is concerned with cost and efficiency in all aspects of production, including equipment, facilities and manpower, and the ways in which they are used. A second area of activity deals with coordination among and control over the actual production processes, including scheduling and other aspects of the dynamic processes of production. The third area, no less important in the long run, deals with construction and maintenance. As we progress, we shall see which of the groups named in Fig. 33 falls into each area.

As we examine these divisions of the corporate manufacturing organization, we shall discuss their functions briefly and also indicate areas of possible employment of chemists and engineers. The general subject we are discussing embraces industrial engineering, business, and management, and is the subject of a number of books (10–15) on production management to which you can turn for more detail in areas of interest.

THE PRODUCTION DEPARTMENT

The responsibility of the production department, briefly, is to provide the proper amounts of acceptable products to meet the actual and anticipated sales commitments of the company. In so doing, it must accept, store, purify, and otherwise prepare raw materials, carry them through the various production processes, and perform whatever finishing operations are required for the products, including packaging and shipping or warehousing. Achieving efficiency in these operations requires careful attention to control and engineering. There are often separate groups for these functions, which work closely with other groups in manufacturing as well as those in other parts of the company.

In small companies, particularly those making quality chemical products commanding high prices, many of the production operations may be carried out directly by chemists and engineers. In larger companies, it is more likely that "blue-collar workers," rather than professionals, will carry out the actual operations while the chemists and engineers have the responsibility for proper control of both products and processes. Employment of chemists and engineers in the production department is higher in the chemicals and allied products industries than in many others because of the highly technical nature of many chemical production processes. It is true, however, that more employment opportunities for chemists and engineers are available in other manufacturing areas. Here are some typical job descriptions within the production department:

Production Supervisor

This man is responsible for the production of a product with acceptable quality and in the amounts desired. He is also responsible for the operating condition of the equipment in his area and particularly for the safety of the men assigned to him. He instructs others in proper operating procedures, follows up to ensure compliance, plans production and maintenance schedules, assists in resolving labor problems, and initiates programs for process improvement to reduce costs.

Production Engineer

The production engineer is responsible for improving the efficiency of the production operations, to create the maximum value from given inputs. He attempts to increase production capacity and avoid bottlenecks. He designs new equipment or reworks existing equipment, installs and starts up new facilities, and improves procedures for materials handling and quality control.

Process Control Engineer

This engineer is responsible for designing and installing new control systems, and improving the scheduling systems coordinating production efforts. He is required to design and develop systems to control many types of machines, processing lines, and materials handling equipment.

THE PLANNING DEPARTMENT

One of the chief functions of the planning department is to provide to production the current and long-range figures for product requirements, as deduced in collaboration with the marketing division. The planning department usually has control over product inventories, and regulates them in accord with established management policies through its link to production control, as indicated in Fig. 33. At the other end of the production line, the planning department has responsibility for raw materials supply and scheduling, as is also indicated by a dashed line in Fig. 33. It is important that these functions are out of the direct line control of the production department, so that no conflicts of interest can arise.

Planning departments are usually staffed by business, accounting, and industrial engineering majors. A few engineers specializing in other areas may be required, but very few chemists are employed directly in such planning functions. In a small company, however, a chemist or engineer frequently assumes responsibility for planning, scheduling, and control of production and raw materials in addition to other duties.

THE QUALITY CONTROL DEPARTMENT

At all stages of the production process, purity and uniformity are important. Every step from incoming raw materials to finished and packaged product must be controlled, checked, tested, assayed. These are the responsibilties of the quality control department. Like planning, the QC department is almost always in a separate line of responsibility from that of the production department, so that its authority to rule on the acceptability of product cannot be overruled by the department producing that product.

Since a relatively small fraction of the output of the chemical industry is in the form of end products sold directly to the consumer, the company you work for usually supplies a product that will be utilized by another company, which will further formulate and fabricate it. The results of these subsequent processes may be greatly dependent on the quality of their raw material, which is the product you produce. If your material changes in properties or goes out of specifications, the result could be disastrous for your customer.

You have a very real responsibility to these customers; a responsibility that goes far beyond good business relations should there be harm resulting from a defect or misrepresentation of the product. You should know that product liability laws, which have been almost universally adopted now (16–18), hold the seller responsible for physical harm resulting to the user or consumer of his product. An illustration of their operation, in which QC was involved, is the following (17)*:

Product Liability: Quality Control

"*Tinnerholm* v. *Parke Davis & Co.*, DC NY 1968. An infant of several months who developed a brain disorder after an injection of Quadrigen, a quadruple antigen with a prophylaxis against diptheria, pertussis, tetanus, and polio, was awarded judgment against the manufacturer for negligence. It was found that the manufacturer *failed to test the products adequately* and failed to warn of inherent risk.

"Settlement: $610,000." (Added emphasis ours.)

Since quality control involves chemical and physical testing of raw materials, intermediates, and products, the QC Department requires the services of chemists and engineers. There are large numbers of job opportunities in QC, especially for those trained in analytical techniques. Here is a typical job description:

Quality Control Chemist

This chemist controls and evaluates all manufacturing materials and operations which affect product quality and reliability. He develops and carries out incoming, in-process, and final inspection procedures; provides for calibration of equipment, instruments, and tooling; develops quality assurance procedures designed to maintain reliability; applies statistical methods to ensure adequate precision and accuracy in sampling and measurement; evaluates product for quality acceptance; and evaluates needs and status of equipment and component reliability.

For further information, a recent book (19) is recommended.

THE MANUFACTURING ENGINEERING AND DEVELOPMENT DEPARTMENT

As Fig. 33 indicates, this department includes a variety of job tasks devoted to what might be termed an R&D function within manufacturing. Most of the employment opportunities for chemists and engineers within the manufacturing division fall in this department. Job titles and functions encompassed within this group or its equivalent may vary widely from company to company, but the following job descriptions are representative.

* Reprinted by permission of *Chemical Technology*.

Product Development Engineer

This development engineer or chemist is responsible for developing modified products and new uses for existing products to meet the needs of customers or the industry. Since his work overlaps that of the new product groups in R&D, he works closely with them, often assuming responsibility for their developments and bringing them to fruition in production. He plans, designs, and tests products to meet performance specifications, and modifies products to obtain greater efficiency, higher quality, and lower costs in their manufacture.

Process Engineer

The process engineer designs new processes, new plants, and changes in existing plants and processes. His objectives are to improve product quality, increase production capacity, and lower costs by eliminating trouble spots and bottlenecks. He develops new processes from the pilot-plant stage to industrial equipment.

The test process engineer plans and writes test specifications and procedures, determines the test equipment required, and devises means for its calibration. His objective is to assure that production and manufacturing schedules are met insofar as test operations are concerned.

Product Engineer

The product engineer assumes responsibility for engineering changes and problems in manufacturing operations, including cost reduction and equipment servicing. He may be assigned to cover the production of a specific product on a continuing basis.

Project Engineer

The project or systems engineer obtains, analyzes, and evaluates technical data to fulfill requirements of the customer and of his own company's facilities. He designs equipment and supervises its installation and initial operation. He may be concerned with new processes and equipment to reduce costs and improve quality, equipment and methods to promote uses and sales, or the provision of technical assistance to plants and customers. He may interpret customer specifications and ensure compliance, provide customer and subcontractor liaison capability, and coordinate efforts of electrical, mechanical, and maintenance groups with those of procurement, production, and manufacturing engineering. He schedules engineering manpower, design, and testing, generates specifications and engineering data for subcontracted items, and sees that all these efforts are coordinated to assure that a properly designed and tested product is delivered to the customer on time.

Manufacturing Engineer

This engineer studies manufacturing processes for possible improvements, prepares cost estimates for new product proposals, develops new standards, and devises means for automating machinery and equipment. His work is closely related to, and he maintains close liaison with, production engineering.

Manufacturing Methods Engineer

This engineer is concerned with the equipment, tools, and instructions for manufacturing a product. He takes drawings and specifications from the design groups and implements their conversion into a working process. His responsibility extends from the time a product has been designed and released for production, through the design, setup, and initial operation of the production line, until it is turned over to production supervisors.

THE PLANT ENGINEERING DEPARTMENT

This department assumes major responsibility for the construction and maintenance of manufacturing facilities:

Plant Engineer

The plant engineer designs and supervises the construction of new plants and facilities, and their development, installation, and maintenance. He is concerned with the layout of plants, machines, and equipment; the provision of electrical and other services; the expansion and remodeling of production lines; and all of the engineering services required in the plant. He also performs economic analysis studies, prepares cost reduction and quality improvement proposals, and participates in engineering and production projects.

TRANSFER OF TECHNOLOGY FROM R&D TO PRODUCTION

In Chap. 6 and earlier in this chapter, we examined parts of the chain of events by which new technology is created and transformed into production processes and products. The actual stage of transition from R&D to manufacturing is somewhat flexible, depending on the nature of the technology and the capabilities of the more applied groups in R&D and the manufacturing engineering and development department.

We have also indicated our feeling that in the future companies will not be able to rely as heavily as they have on a continuing stream of new products to maintain their growth in productivity and profits. Product and process improvement will become increasingly important, we expect, as the chemical industry responds to the challenges of the future. The pressures on manufacturing engineering and development—and in fact on all of manufacturing—will increase. An atmosphere conducive to innovation will have to be established throughout the organization if productivity and profits are to improve.

As an aside, we note that the effect of increased job pressures on the individual may not be all bad. "Contrary to popular opinion, job pressures are not necessarily undesirable; in fact, certain pressures seem to enhance researchers' job attitudes and importance . . . research pressures are most likely to be effective when they reflect internal commitment, personal control, direct task demands, a history of success, and organizational support" (20).

In addition to the transfer of technology from R&D to manufacturing engineering and development, there are two stages which are critical to getting the new product or process into fullscale production: the scaleup and pilot-plant stage, and the production startup. We do not intend to discuss these parts of the product life cycle, but refer you instead to several recent references (21–24). Of course, it is in the best interests of both management and employees to minimize both the time and the cost of these stages. Good communications at all levels is, as always, important for achieving these results.

COSTS AND PROFITS

COSTS

Although we saw earlier that the American chemical industry is made up of a mixture of large dedicated plants, general-purpose synthesis plants, and formulating plants, it is the huge output of the dedicated plants that characterizes the industry as a whole. In terms of overall output, the chemical industry is largely a continuous-process industry that requires a very heavy initial investment.

As a result, the cost of making products in the chemical industry is distributed rather differently than in manufacturing industry as a whole. In 1973, the chemicals and allied products industries accounted for 11.7% of the new capital investment for all manufacturing, but only 5.2% of the total employment (1). And the average invested capital for plants and equipment was almost $28,000 per employee in the chemicals and allied products industries, almost twice as much as for all manufacturing. As a result of this higher capital investment expenditure, with correspondingly higher depreciation and taxes, the fraction of product cost resulting from labor and materials is lower in the chemicals and allied products industries. Exact figures are hard to come by, but in 1965, the relative cost of labor in the chemicals industry was only 75%, and that for materials 82%, of that for all manufacturing (3).

These higher capital costs reflect the large capacity required to produce at maximum efficiency for many chemicals and the intricate nature of the equipment involved. Maintenance and replacement costs are correspondingly high, as shown by current economic indicators (25) which seem to rise exponentially in these inflationary times.

One consequence of the fact that the chemical industry is largely devoted to continuous processes is that added output usually involves relatively small incremental costs. Thus, the cost per unit of product decreases as volume increases, and a lower price can be used as an incentive to increase sales. In fact, in some segments of the industry the prices of a wide variety of products are related to their current individual production volumes in a very regular way; Fig. 34 provides an example.

A number of recent articles (26–30) and a book (31) deal with costs, cost estimation, and cost control in the chemical industry.

PROFITS

"Profits are the vital nutrient feeding the growth and building the future well-being of the chemical industry" (1). In commenting on profits, we limit our remarks to the period ending around the middle of 1974, in which the economic climate was, in comparison to that which followed, still reasonably settled.

Recovering from the mild recession of 1970, the chemical industry did very well in the next two years, with earnings advancing 23% in 1972 and 41% in 1973. In 1973, basic chemical producers earned 6.5 cents per dollar of sales, up from 5 cents in 1970–1971 but well below the level of 8–9 cents of the mid-1960s (1).

Profits and earnings are affected by many factors, including R&D expenses, raw material and labor costs, productivity, sales prices and volumes, accounting practices with respect to inventory and depreciation,

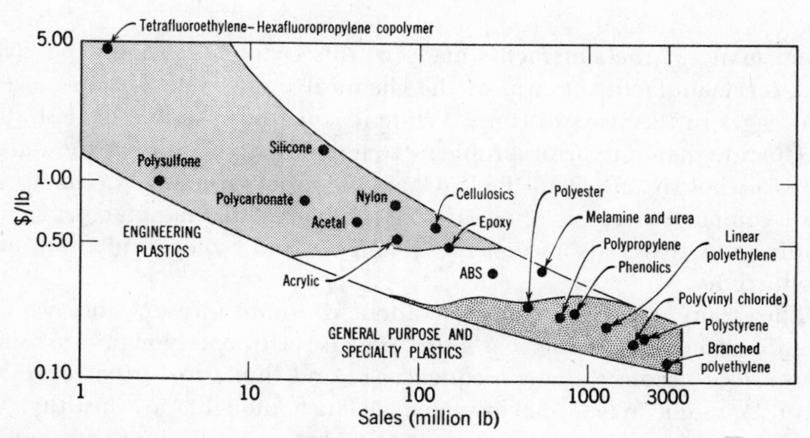

FIGURE 34. Relation between prices and production volumes for a variety of plastics (6). (Reprinted by permission of John Wiley & Sons, Inc.)

taxes, and many others. Since they represent only a small fraction of income, profits can fluctuate widely with small changes in some of the other factors. Changes in sales volume are a major factor in determining profit levels in any year, as can be seen by comparing the trends in the two between good and bad years. Any deviations from the parallelism of these two can usually be explained in terms of other economic factors, such as rising costs or declining prices.

Over a period beginning well before World War II, the profit levels in the chemical industry were generally high (except during the war years) but have rather steadily declined over the years as the effects of more intensive competition, both domestic and foreign, excess capacity, and declining prices were only partially offset by increased volume, the introduction of new products whose early profit levels were high, and lower raw material costs. On average, profits were in the range of 14% of the sales dollar in the late 1930s, near 10% around 1950, when the effects of the war were no longer felt, 8% in the early 1960s, and as we saw above, near 5% in 1970–1971. Whether the 1972–1973 recovery signals a reversal of this trend, only time will tell. In comparison to all manufacturing, the chemical industry has traditionally had higher profit levels: Only in five years since 1925 has the profit margin for chemicals been below the industry average. Reference 32 gives a good historical survey of profits in the chemical industry, and the annual "Facts and Figures" summaries (1) in *Chemical & Engineering News* will keep you up-to-date.

CHALLENGES AHEAD

From many of the statements made in this chapter, it should be clear that the manufacturing arm of the chemical industry faces some severe challenges in the years to come. While it is neither possible or desirable to separate manufacturing problems completely from those of the industry as a whole (Chap. 2), of R&D (Chap. 6), or of marketing (Chap. 8), we shall comment here on some critical challenges for manufacturing, including raw materials and energy, plant size and ecology, inflation, and productivity.

The steady growth of industry cannot continue forever, since we are a finite population living in a finite universe. Ultimately man must start to consider seriously the consequences of and limits to further development. A recent article (33) comments: "Such thoughts are healthy; we are fortunate that humanity does not wish to rush blindly to its destiny as lemmings do." It would seem that in the highly industrialized countries we have now reached that point in time. We must ask how much longer the expansion of the chemical industry can and should be extended, and how this expansion can be channeled and directed so as to bring benefits to all.

ENERGY AND RAW MATERIALS

We deliberately treat these two topics together because they are in fact inseparable.

Chemical raw materials can be divided into two types (33): those that are used almost exclusively by the chemical industry, such as sulfur, phosphate ores, fluorspar, or salt; and those, such as petroleum, that are used by several branches of the economy. Many of these are primarily energy sources and only secondarily chemical raw materials. On the other hand, such sources of energy, including electricity, are just as much raw materials essential to the operation of a chemical plant as any other.

Raw materials can be classified (34) as renewable (plant- or animal-based), fossil, and inorganic. Regardless of type, continuity of supply and future availability with only gradual changes in prices are the major concerns that must be faced. To stave off ultimate, irrevocable shortages, the industry must consider the approaches of avoiding waste, changing the balance of sources to favor renewable ones, and introducing new technologies directed at better utilization of low-grade sources (e.g., shale oil instead of petroleum) and recycling.

All of these considerations are inextricably intertwined with energy, and much has been written about the supply of and demand for energy in the chemical industry; we cite only two recent articles (35, 36) indicative of the conflicts that abound in this area. In agreement with another author (37), we note that it takes energy to get energy, and it may well be that we have already reached the point of diminishing returns in efforts to utilize alternative sources such as oil from shale. In any case, it seems both reasonable and necessary to consider the net return of energy over the amount of energy used to develop new sources.

PLANT SIZE AND ECOLOGY

Although we saw earlier that the most economical size for a dedicated, continuous process chemical plant is very large, there are reasons to think that practical limits on plant size may have been reached (33). Some of these reasons are technological: limits on permissible quantities of effluents or other substances that have environmental impacts, problems of startup, loss of flexibility, difficulties in substitution of production in case of plant failure or shutdown. Other reasons are based on human or moral considerations: Problems of information, motivation, and personnel relationships—in short, difficulties in managing—arise if plants exceed a certain size. So does impact on the environment from the human standpoint, especially in densely populated areas. A dilemma therefore arises: The chemical industry must remain profitable to survive, and in so doing must increase productivity, by increases in plant size among other means. On the other hand, increased plant size can no longer be assumed to be a possible alternative.

INFLATION

As we write in the dark days of double-digit inflation, there seems little we can add to what is well known, even if we had far more expertise than we do in the field of economics. The impact of inflation on the chemical industry is no different than on any other, and we can only note the seriousness of the problem. A recent article (38) is pertinent to the chemical industry in its global aspects.

PRODUCTIVITY

In mathematical terms, productivity is simply a ratio of the output of an industry to some input. We are concerned with the ratio of the output of the chemical industry, measured by sales dollars or pounds of product, to the number of workers involved or the number of man-hours worked. Productivity in the chemicals and allied products industry has traditionally increased more rapidly than that for all manufacturing industries (Fig. 35) but the two curves appeared to meet for the first time in 1973, as labor costs increased significantly. In addition, the growth of productivity of professionals in the chemical industry lagged badly behind that of production workers (39). Has the chemical industry reached a crossroads, a point of no return?

It is clear that if the challenges of the future are to be met, the chemical industry must improve its productivity at all levels, and that this requires attention to human relations as much as to material processes and products. In a recent speech (40) Charles B. McCoy, then president and board chairman of Du Pont, "boil[ed] the productivity issue down to three conclusions:

> First, output per man-hour is not high enough in industry or elsewhere. If the nation is to meet its environmental and other goals, productivity has to go up. . . .
>
> Second, the focal point for higher productivity has to be the private

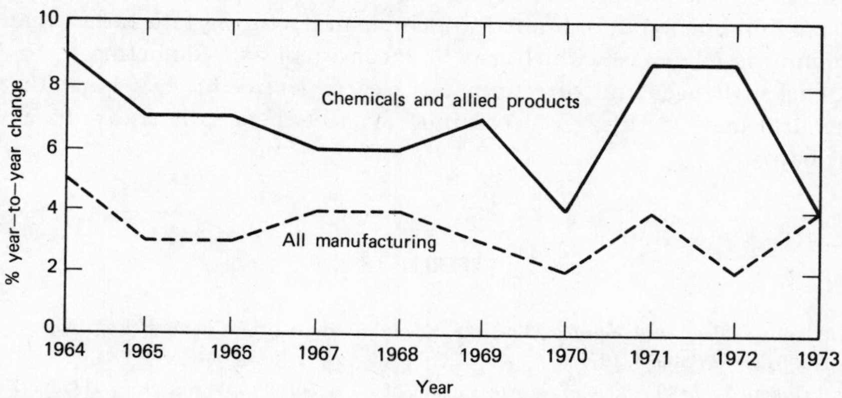

FIGURE 35. Rates of increase of productivity for the chemicals and allied products industry and for all manufacturing in recent years (1). (Reprinted by permission of *Chemical & Engineering News.*)

sector rather than government, and a healthy part of the job has to be done in the manufacturing and processing industries.

Third, those of us who work in industry or who are concerned with its problems find a large part of the responsibility sitting on our own doorsteps. . . ."

McCoy then expressed his advice in terms of four areas that concern him:

- A new national attitude is needed, in which productivity is looked at the way safety is in industry, as an everyday part of every individual's job. We must abandon the idea that productivity is what the other person ought to deliver.
- More personal initiative must be shown in tackling the problem. The top brass can't order increases in productivity.
- We must realize, and act accordingly, that it isn't necessary to spend lots of money to increase productivity. The objective is to save, not spend.
- The more authority and volition is given a person, the better the chance he will increase his productivity. We should err, if at all, in the direction of overestimating rather than underestimating people.

In a somewhat more materialistic approach (41), a Harvard professor of business administration suggests that we see the problem not as "How can we increase productivity?" but "How can we compete?" He feels that the problem encompasses the efficiency of the entire manufacturing organization, not just the work force, and recommends that each plant be focused on one explicit, limited, concise manufacturing task rather than attempting many tasks which may be inconsistent and conflicting.

The challenges and directions are clearly set forth; let's hope that chemical manufacturing can continue to meet them as it always has in the past.

REFERENCES

1. Anon., "Facts and Figures: The U.S. Chemical Industry," *Chem. & Eng. News* **52** (22), 21–52 (June 3, 1974).
2. *Chemistry in the Economy*, American Chemical Society, Washington, D.C., 1973.
3. Jules Backman, *The Economics of the Chemical Industry,* Manufacturing Chemists Association, Washington, D.C., 1970.
4. Patricia Noble, ed., *Kline Guide to the Chemical Industry,* 2nd ed., Charles H. Kline, Fairfield, N.J., 1974.
5. B. G. Rueben and M. L. Burstall, *The Chemical Economy,* Longman's, London, 1973.
6. Fred W. Billmeyer, Jr., *Textbook of Polymer Science,* 2nd ed., Wiley-Interscience, New York, 1971, Chap. 7.
7. Anon., "Chemicals Output Starts to Plateau," *Chem. & Eng. News* **52** (4), 11–12 (Jan. 28, 1974).
8. William F. Falwell, "Chemical Capital Spending Boom Continues," *Chem. & Eng. News* **52** (48), 7–8, (Dec. 2, 1974).
9. Harold Stieglitz and Allen R. Janger, *Top Management Organization in Divisionalized Companies,* Personnel Policy Study No. 195, National Industrial Conference Board, New York, 1965, pp. 85–90.
10. Elwood S. Buffa, *Modern Production Management,* Wiley, New York, 1973.
11. Joseph A. Litterer, *Organizations: Structure and Behavior,* 2nd ed., vols. 1 and 2, Wiley, New York, 1969.
12. Joseph A. Litterer, *The Analysis of Organizations,* vol. 2, Wiley, New York, 1973.
13. Edwin S. Roscoe and D. G. Freark, *Organization for Production,* 5th ed., Irwin, Homewood, Ill., 1971.
14. Gordon B. Carson, *Production Handbook,* 3rd ed., Ronald, New York, 1972.
15. Donald E. Ramlow and Eugene H. Wall, *Production Planning and Control,* Prentice-Hall, Englewood Cliffs, N.J., 1967.
16. Henry R. Piehler, Aaron D. Twerski, Alvin S. Weinstein, and William A. Donaher, "Product Liability and the Technical Expert," *Science* **186**, 1090–1093 (1974).
17. George W. Woods, "Production Management and Liability Law," *Chemtech* **1,** 310–313 (1971).
18. Harry M. Philo, "Product Liability $\gtrless$ Product Safety," *Chemtech* **1,** 88–92 (1971).
19. Elwood G. Kirkpatrick, *Quality Control for Managers and Engineers,* Wiley, New York, 1970.

20. Douglas T. Hall and Edward E. Lawler, III, "Job Pressures and Research Performance," *Amer. Sci.* **59**, 64–73 (1971).
21. V. T. May, "Process Development Costs & Experience," *Chem. Eng. Prog.* **69** (2), 71–75 (1973).
22. Ernest O. Ohsol, "What Does It Cost to Pilot a Process?" *Chem. Eng. Prog.* **64** (4), 17–20 (1973).
23. F. R. Bradbury, "Scale-up in Practice," *Chemtech* **3**, 532–536 (1973).
24. Joseph Mizrahi, "People, Organization, and Process Implementation," *Chemtech* **2**, 459–464 (1972).
25. "Economic Indicators," published in each issue of *Chemical Engineering.*
26. John W. Lunger, "Cost Control Through Scope Control," *Chemtech* **4**, 413–417 (1974).
27. Anon., "Plant Costs Zoom," *Chem. Eng.* **81** (13), 136–137 (June 24, 1974).
28. R. W. Newkirk, "Machinery Costs *vs.* Energy Conservation," *Chem. Eng.* **81** (13), 138–140 (June 24, 1974).
29. N. S. Eastman, "Cost-Improvement Techniques Spur Widespread Savings," *Chem. Eng.* **80** (28), 102–112 (Dec. 10, 1973).
30. F. A. Holland, F. A. Watson, and J. K. Wilkinson, "Manufacturing Costs and How to Estimate Them," *Chem. Eng.* **81** (8), 91–96 (April 15, 1974).
31. Herbert Popper and the Staff of *Chemical Engineering*, eds., *Modern Cost-Engineering Techniques*, McGraw-Hill, New York, 1970.
32. Ref. 3, Chap. 10.
33. M. Jean Montet, "Problems of the Chemical Industry Today," *Chemtech* **4**, 268–272 (1974).
34. Duncan S. Davies, "Raw Materials for Chemicals," *Chemtech* **4**, 135–139 (1974).
35. Milton F. Searl and Sam H. Schurr, "An Overview of Supply/Demand for the Next Decade," *Chem. Eng. Prog.* **69** (6), 27–34 (1973).
36. Lena C. Gibney, "Energy Report Urges Shift to Electricity," *Chem. & Eng. News* **52** (47), 12–13 (Nov. 25, 1974).
37. Wilson Clark, "It Takes Energy to Get Energy: The Law of Diminishing Returns Is in Effect," *Smithsonian* **5** (9), 84–90 (Dec. 1974).
38. Paul A. Samuelson, "The New Bias Toward Inflation," *Chemtech* **4**, 656–659 (1974).
39. Michael Heylin, "White-Collar Productivity Troubles Industry," *Chem. & Eng. News* **50** (30), 6–7 (July 24, 1972).
40. Charles B. McCoy, "Productivity: Myths and Missions," *Chem. Eng. Prog.* **69** (5), 37–41 (1973).
41. Wickham Skinner, "The Focused Factory," *Harvard Bus. Rev.* **52** (3), 113–121 (May–June 1974).

CHAPTER 8

MARKETING

If you were not already familiar with the terminology, you may have wondered why we have used the term *sales* at some times and *marketing* at others for that part of industry responsible for getting the finished product from the factory to the customer. Is there a difference, and if so, what are the definitions of the terms? Let's answer with a couple of quotations:

> Marketing is the business of moving goods from the producer to the consumer. It is not synonymous with selling, although selling is the heart of it, because [marketing] also includes such functions as product development, testing, pricing, distribution and advertising (1).
>
> While sales is only the consummation of a transaction, marketing encompasses *all* of the analytical, promotional, and service functions that permit these transactions to be initiated and continued on an increasingly profitable basis (2).

The second author goes on to provide a functional definition of marketing as "determining what to sell to whom at a profit; via what terms, conditions, and channels of distribution; and creating and managing programs to generate service, and expand these sales"; for an illustration of this definition see Fig. 36.

MARKETING APPROACHES

There are several approaches to marketing, which we mention briefly before launching into a description of the marketing division of a company and its functions.

INDUSTRIAL VS. CONSUMER MARKETING

The distinction between industrial and consumer marketing (3) corresponds to that made in Chap. 7 between the manufacture of basic chem-

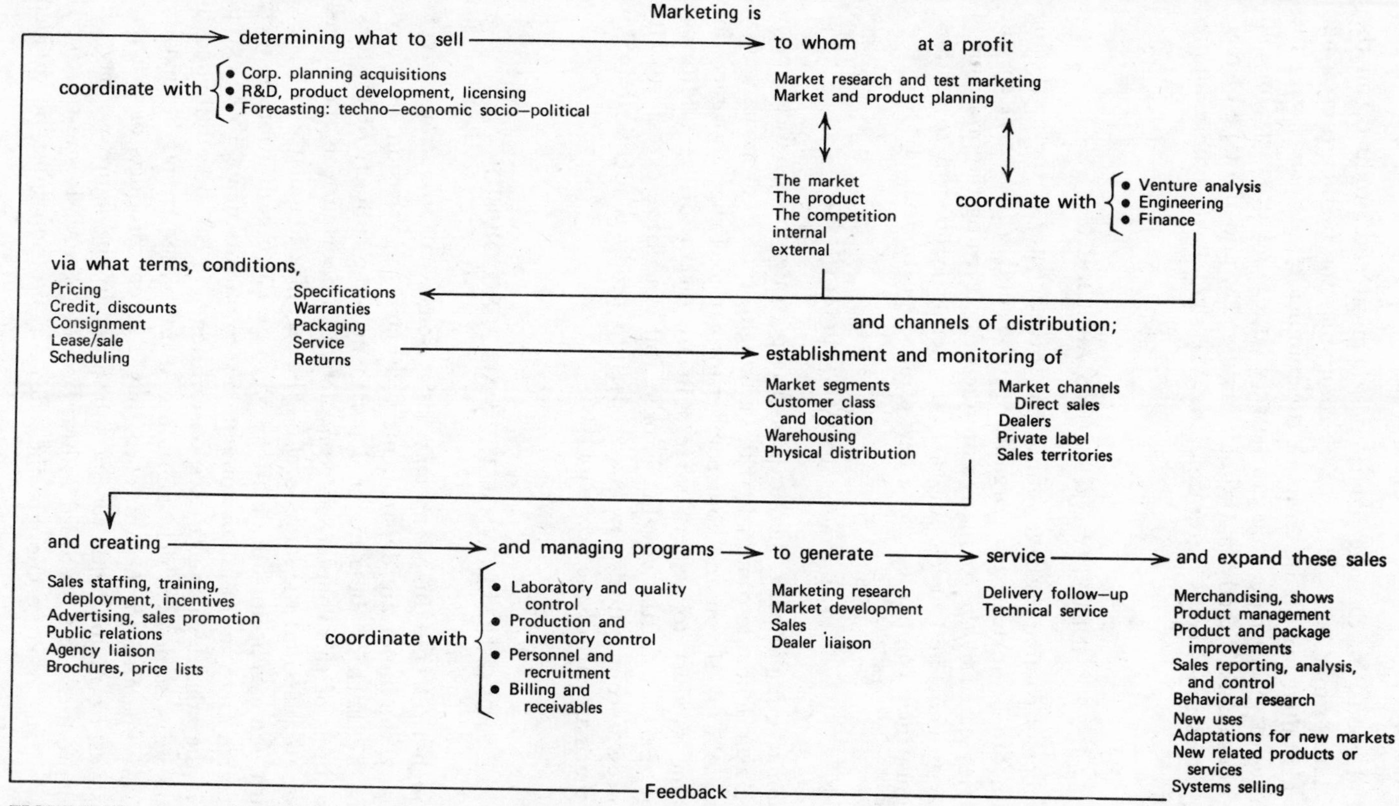

FIGURE 36. A functional definition of marketing (2). (Reprinted by permission of *Chemical Technology*.)

icals and goods, on the one hand, and chemical end products, on the other. The former are sold to other companies for further processing; this is a limited market in terms of the number of customers, and the quantities of product per sale can be very large. End products are sold to the consumers, directly or through middlemen, and the type of marketing effort required is quite different. We shall have some examples later on.

INDUSTRY OR SYSTEMS APPROACH

A number of companies have found that the efficiency of the marketing division can be increased by organizing it around segments of specific industries served by the company's products. By learning the problems of such a segment, and developing and marketing systems to solve those problems, these companies create new profit opportunities (4).

THE COMMODITY APPROACH

In the marketing of basic chemicals on a large scale, the product is often viewed as a definable commodity, to be bought, traded, or exchanged as well as sold (6). If you and your competitor can exchange customers for the same product to achieve savings in freight or utilize additional storage space to mutual advantage, or if you can utilize excess capacity to process a competitor's raw materials into products identical to his own, that's all part of the new game.

ORGANIZATION OF THE MARKETING DIVISION

Although titles and functions may differ somewhat from company to company, the important departments of a marketing division are shown in Fig. 37. Each of these major departments—the subjects of the next five sections of this chapter—is equivalent in importance to the others, so they are placed at the same level on the organization chart.

With one exception, the details of the internal organization of these departments are not of major consequence to our discussion. The exception is the sales department. Responsibilities here can be subdivided in various ways, some of which are indicated in Fig. 38. It may be that the company operates on the approach that different products or different industries are major fields for specialization; in the figure we have selected products, but the word industry could be directly substituted. If the company is large enough, the sales organization must be divided

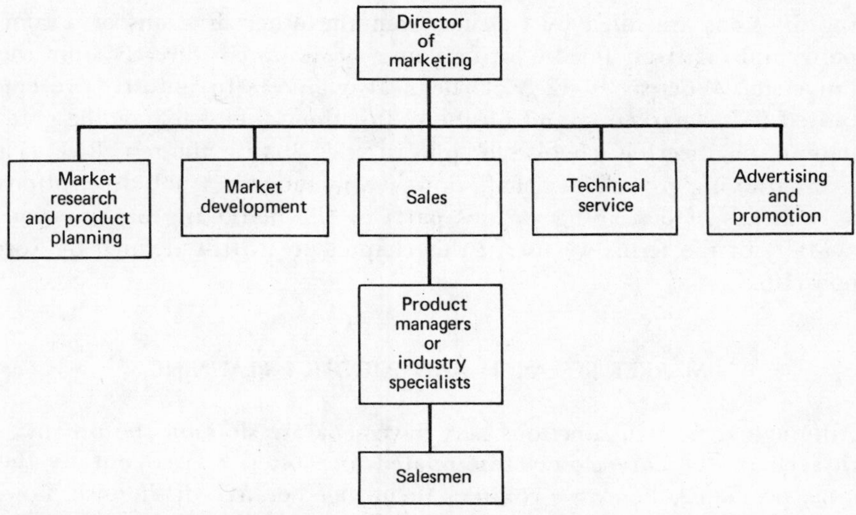

FIGURE 37. Typical organization chart for a marketing division.

further by district or territory, including international sales. Another alternative is to subdivide by geographical areas first, and by products or industries at a lower level, if at all. These various organizational alternatives are discussed in refs. 7 and 8.

Before we describe the various departments of a marketing division, a comment on job opportunities is pertinent. Most positions in market-

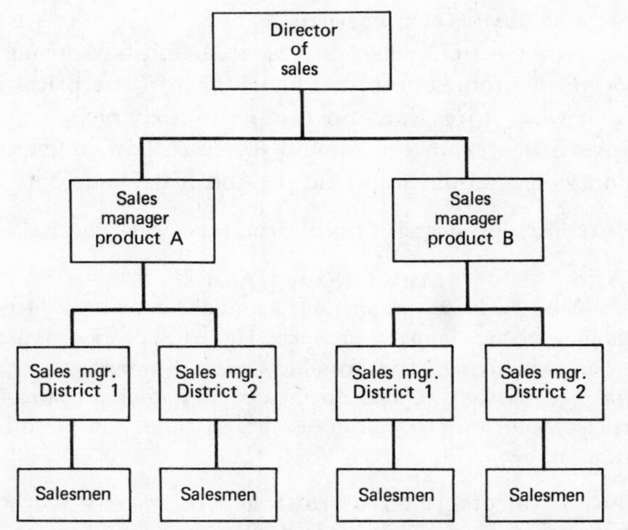

FIGURE 38. Sales department organization by product specialty and territory.

ing divisions are filled by transfer from the other divisions of a company, and relatively few by hiring young professionals directly from the university. Whereas 10–12% of chemical engineers in industry are employed in a "marketing and business" function, only 4–5% of the graduates of engineering schools seek jobs in sales (9). For the rest, R&D is a likely training ground. We shall, however, include a few job descriptions as a means of describing various parts of the marketing organization.

Many of the terms we use in this chapter are wittily defined by Zornow (10).

MARKET RESEARCH AND PRODUCT PLANNING

Although these two functions may have separate slots on the organization chart, they are closely interrelated and often carried out by the same personnel; hence we consider them together. We distinguish, however, between market *research,* which deals mainly with exploring the possibilities of new products, and market *development,* which has responsibility for the markets for existing products throughout their life cycle. Market development is considered in the following section.

Berenson (7) defines market research as the task of gathering and analyzing information about marketing problems and objectives. Market researchers provide the answers to such questions as:

- Who uses our product?
- Who are our customers, where are they located, what are their social, economic, and buying characteristics?
- Who is our competition, where is it located, and how strong is it?
- What sort of a product line should we have, what prices should we charge, and what prices does our competition charge?
- What advertising campaign should we use, how much should we spend, and where should we spend it? And many more.

Here is a typical job description of a market researcher:

Market Research Analyst

The market analyst evaluates the market potential for new products, as well as new markets for existing company products. He maintains an awareness of current markets and marketing trends, conducts research programs to study market potential and new markets, reports on these studies, and prepares information on and participates in contract and license negotiations, and possibly in merger and acquisition studies.

These duties and others related in a general way to market research are described in references 1, 7, 11, and 12.

NEW PRODUCT PLANNING

In Chap. 6, we followed the life cycle of a product from the standpoint of R&D, and in the next section of this chapter we review the contributions of marketing to this cycle. However, in both instances we assumed that there was a product, fully born and ready to enter into the cycle. But products don't just materialize; there are several stages through which ideas and concepts must move in order to develop a product worth commercialization. The role of market research in new product planning is to determine the product fields of primary interest to the company, establish a program for planned idea generation, and collect ideas for potential products in an organized way. These ideas are then expanded into product proposals, facts and opinions are gathered, and the first of a series of screening steps in which the ideas are appraised is undertaken. Those that survive this stage are subjected to business analysis to determine the desirable market features for the product. R&D is consulted so that technical and business feasibility and specifications can be developed simultaneously. It is to this point we shall return in the next section. References 7 and 14–16 amplify and provide examples in the area of new product planning.

SALES FORECASTING

Another function of marketing research and planning is to estimate the volume of future sales, a task requiring not only a knowledge of economic conditions and a broad background in marketing, but the ability to make a realistic appraisal—in today's world—and to review and revise it at frequent intervals. Reference 17 provides further information.

OTHER FUNCTIONS

Reference 7 provides detail on the following additional functions and responsibilities of marketing research and planning;

- The marketing plan, a document specifying a systematic program for achieving marketing goals according to a time schedule. It details what is to be done and how, by whom, and by when.
- The marketing audit, a systematic analysis of the company's entire marketing program, including policies, personnel, functions, resources, and structure.
- The interface between marketing and R&D—although we shall now see that market development continues and expands this relationship.

MARKET DEVELOPMENT

Market development can be defined (18, 19) as the interface between the creators of a new product—jointly R&D, manufacturing development, and market research and planning—and the customers for that product. More than any other group in marketing, development has the responsibility of guiding the life cycle of a product from commercialization through abandonment. In so doing, it relates with the other groups, such as market research, technical services, advertising, and sales. While there is no one "best way" for organizing and carrying out market development, for every case must be considered according to its own needs, the need for this function is continuous from the time the product is introduced in the marketplace until and even past the time production has ceased.

THE PRODUCT LIFE CYCLE

Figure 39 presents a market development-oriented view of the product life cycle, this time omitting the early stages prior to commercialization shown in Fig. 19, and providing detail of the later stages of the cycle. We now describe the characteristics of these stages, and the role that market development plays in each of them, complementing the responsibilities of R&D and manufacturing in the same stages as outlined in Chaps. 6 and 7, respectively.

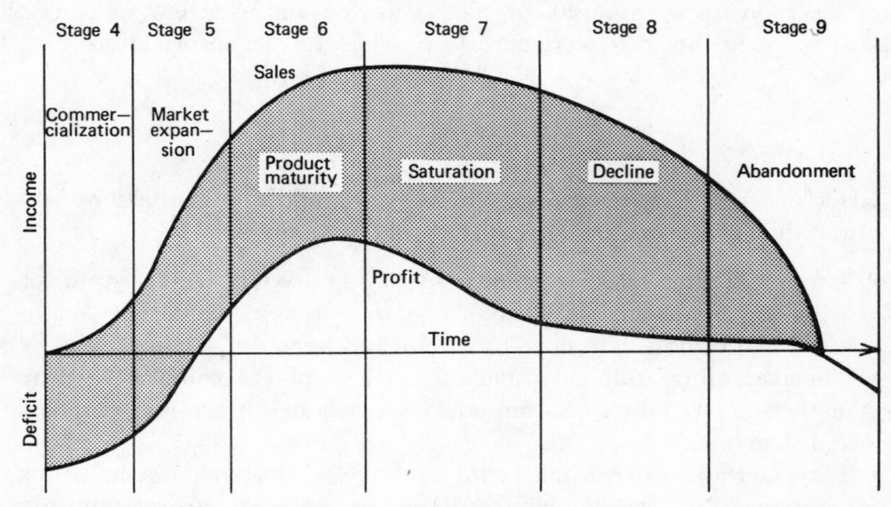

FIGURE 39. The product life cycle, with emphasis on the later stages (3). See also Fig. 19 and the discussion in Chaps. 6 and 7.

In stage 4, *commercialization,* the new product evolved in R&D with the assistance of the new product planning groups in market research is launched into the commercial market. There is a slow rise in sales volume, and the deficit incurred in the earlier stages—conception, demonstration of feasibility, and producted development—begins to decrease as some profits are earned. This is still a time for caution: The product position is weak, with high production and marketing costs and the need for much technical service. At this stage market development must work out the basic marketing and pricing policies for the product, planning for broadening distribution as production evolves from pilot-plant to full-scale, for broadening and changing of the product line as innovation proceeds, and for broadening of competition as times goes on. Prices are comparatively high at this stage and advertising is of the informative kind.

This introductory stage of the life cycle is difficult to control. Some products catch on quickly, while others grow very slowly and may undergo a period of decline in sales before "taking off" into the next stage of growth. Of course, some products die here, too. By definition, the deficit incurred by the product is not converted into a profit until the next stage is reached.

Stage 5 is the phase of major growth or *market expansion* of the product. It is heralded by a rapid rise in the sales curve. If this rise results from intensive (and expensive) advertising and promotion, as is often the case, the profit curve may not cross from the red into the black until the product is well into this stage. It is here that the market development team must put into effect the distribution and pricing policies which, if properly selected, will permit the growth of sales and profits to continue. The introductory prices may have been set high enough to limit the deficit incurred in developing the product. They must now be reduced; if they are set too low, a profit may not be realized even though sales volume is high, but if they are set too high, the product may not be attractive to customers and additional competitors may enter the market.

When the rate of growth of sales begins to slacken, and profits start to drop, product *maturity* (stage 6) has been reached. The markets have not yet been satisfied; sales can be increased by additional merchandising, but the costs of this outweigh the profits that result. Competition has increased. Marketing development looks ahead, making long-range plans for the inevitable decline that will lead to the abandonment of the product. One obvious step for the company is to introduce new and improved products at or before this stage so that their increasing profits will compensate for the declining profits of the more mature products whose history we are considering.

Maturity is followed by *saturation* (stage 7) when it is no longer possible to increase sales despite strenuous efforts. There may be a decline in sales, or the highest level may be held for a long period, but profits decline because of intense competition in this phase. R&D and technical service are reduced, small (less profitable) orders are refused, the product line is simplified, advertising uses the "hard sell," prices may be cut, but marketing costs increase.

The inevitable *decline* (stage 8) follows. There is still a profit to be made, but both it and sales are declining, and nothing can be done to halt the drop. Smart marketing management will recognize and prepare for the death of the product. Advertising is cut off, plant size is decreased and equipment is salvaged for use elsewhere, first-class personnel are transferred, the number of distributors may be cut, the product line is slashed.

The final phase of *abandonment* of the product (stage 9) is one in which customers are advised, and hopefully weaned away to other company products in the introductory or growth stages. Sales cease, plants are closed or shifted to alternate missions. Unfortunately, marketing costs continue to be incurred long after these events, since the product which is still in the hands of the customer must be served. Furthermore, the responsible company must consider its customer's continuing needs. This may lead to sale of the technology to another producer, or delay of the abandonment for some years. Social responsibility, to workers, customers, and society, cannot be overlooked. Abandoning a product can be at least as difficult as introducing it.

This review of marketing development responsibilities in the product life cycle was drawn from refs. 3 and 7.

SALES

As the level of technology in the American chemical industry has increased, and as the systems approach to marketing has become more prominent, the scientific and engineering training required in the sales group of marketing has also increased. Even the so-called line salesman, who actually calls on the customers, is now technically competent. He is backed up by a staff of sales engineers or sales technologists who are as fully professional as any scientist or engineer—though their professional status is not yet as well recognized (20). We offer the following job descriptions, drawn in part from ref. 7, to describe typical functions of the sales department:

Sales Technologist

The sales technologist is responsible for customer contact and technical specification analysis. This includes preparing quotations and initiating and recommending product or design changes. He may also advise on application problems, handle correspondence on product quality or toxicity, visit customers with salesmen on serious problems, and prepare evaluation reports on experimental products.

Application Engineer

The applications engineer (or scientist) provides specialized support for the salesmen. He gathers and analyzes information, answers inquiries from the field, and devises product modifications and special designs to meet the customer's needs. He keeps abreast of trends in product design and competitive situations. He is an authority on various types of products and their applications, and a counselor to the field sales force. He often makes technical calls on customers.

Field Service Representative

The field service representative is in close contact with the customer. His duties include supervising installation and instructing the customer's personnel in the operation of equipment and the use of the product.

Sales Engineer

The sales engineer solicits business from a specified industry or for a specific group of products. He makes regular sales calls on existing customers, solicits new accounts, plans sales itineraries, watches for and reports competitor activity, keeps well informed on all new or modified applications of company products, and performs a variety of reporting, correspondence, and promotional activities.

Product Manager

The product manager has general responsibility for a line of products. He reviews and recommends advertising, recommends changes in product type or quality, advises on how the company should conduct its marketing for maximum efficiency, and is fully informed on costs. He visits key customers with the salesmen, evaluates sales performance and recommends changes, assembles intelligence from the field, and establishes and controls inventories. He keeps current on competitive activity, initiates requests for marketing research, seeks out new applications for his products, coordinates production schedules with manufacturing, determines prices, and maintains sales and profit records. He is the basic planner for his product line, and the tie line for it between marketing and management.

INTERNATIONAL SALES

In the postwar market up to the early 1960s, export sales were a small part of the output of the American chemical industry, in a seller's market. An export department or international division sold to local sales agents abroad, who in turn sold to customers in their own countries. Since that time there has been a change in international sales operations as a whole, as many companies have established their own sales agents, sales offices, and subsidiary companies around the world. The international division has become just another area division within sales, equivalent to the several domestic divisions which may exist, and the product manager has global instead of just local responsibility (21). In some instances, the multinational companies described in Chap. 2 have evolved.

TECHNICAL SERVICE

Technical service may be defined (3) as any service performed by the seller to enable the customer to use the product more effectively. The American chemical industry has always placed great emphasis on providing this service, whose nature can range from a brochure describing the properties and uses of the product to technically trained personnel spending months in the customer's plant solving his problems in the use of the product.

As materials, applications, and equipment have become more sophisticated, the nature of technical service has changed (22). Some of the questions that a technical service group must face these days are:

- How can you predict the service life of a plastic over a 20- to 30-year span?
- How can you predict the long-range hazard to the population of a new soap bacteriostat or food additive?
- How can you deal with subjective properties of products such as taste and odor, color, or "feel," "handle," and "drape" in a fabric?
- How do you wean "old timer" color matchers away from visual techniques to the use of instruments and computers?

From some of these challenging questions, it is clear that government regulation has a strong effect on the nature of technical service for today's products.

Some of the technical services commonly furnished in the chemical industry are (3):

- Develop new products on customer's requests.
- Develop new uses for existing products.
- Introduce products to the customer in either his plant or yours.
- Write product technical bulletins.
- Train sales personnel and advise field salesmen.

- Handle customer inquiries, providing a broad range of assistance.
- Train customer personnel.

Is all this service necessary? Yes, either because of competitive pressure, or because some products are just too difficult for untrained people to handle. And it is often good defensive strategy to supply technical service even if the customer didn't (didn't know enough to?) ask for it, to make certain that the product is used correctly.

But technical service can be very expensive, and if its costs are not carefully controlled, they can eat significantly into profits.

ADVERTISING AND PROMOTION

Advertising is essentially a device to communicate with a large number of people at low cost. The basic function of the advertising and promotion group in marketing is to create an awareness of the company and its products within the industry, in order to maximize the effectiveness of all the marketing activities. This group must be both creative and administrative. It must foster imaginative programs and presentations, control the production of promotional materials, and administer a relatively large budget. It must interact with advertising agencies and suppliers in planning selling programs. Clearly, the members of this group must understand the needs and problems of customers and salesmen.

Some of the functions of the advertising and promotion group are (7):

- Establish and regulate the annual programs of advertising, sales promotion, and trade-show participation.
- Negotiate with the advertising agency, publishers, printers, artists, and other sources of supply and services, within a prescribed budget.
- Originate and supervise the production of service literature, catalogs, data sheets, and technical bulletins.
- Coordinate promotional publicity with the company's public relations staff to generate articles based on the company's products.

References 23 and 24 provide additional information on the role of advertising in the chemical industry.

PRICING

There is a good deal of strategy involved in determining the prices of chemical products, and the subject cannot be fully separated from related parts of the marketing mix, such as market research, product planning, advertising, and technical service. In outlining some of the basics of pricing policy, we follow an excellent section in ref. 7; refs. 17 and 25 have more information.

Unlike the marketing functions described so far, each associated with a single group, pricing can be carried out by a variety of personnel, not all of whom are even in marketing, as indicated in Table 23. Among the factors that these pricing personnel have to consider are:

- The product's cost, which sets an obvious long-term limit to the price.
- The product's value, that is, how well it satisfies certain needs.
- The way the product's sales change as its price is changed. In the usual case, sales go up as prices go down, but the opposite might be true, for example, for a perfume.
- The company's goal for return on investment.
- The company's market position, encompassing such factors as location or service policy which might influence the price a customer is willing to pay.
- The nature and extent of competition.

BREAK-EVEN ANALYSIS

The break-even point, at which total revenue equals total cost, is an important factor in determining prices. The cost of a product is the sum of fixed costs (such as executive's salaries, insurance, and rent) and variable costs (those proportional to production, such as raw material, direct labor, and energy costs). The relations among these are shown in a break-even chart, Fig. 40. Since total revenue is the product of price and

TABLE 23

Personnel Responsible for Determining Prices*

Company	Personnel
American Can	President
U.S. Steel	Executive vice-president
Johns-Manville	Vice-president
General Electric	Department heads
General Foods	Division managers
Du Pont	District sales manager and division sales director
Alcoa	Product sales manager
Sears, Roebuck	Individual buyers (hundreds of them)

* Adapted from (7). Reprinted by permission of author and the American Chemical Society.

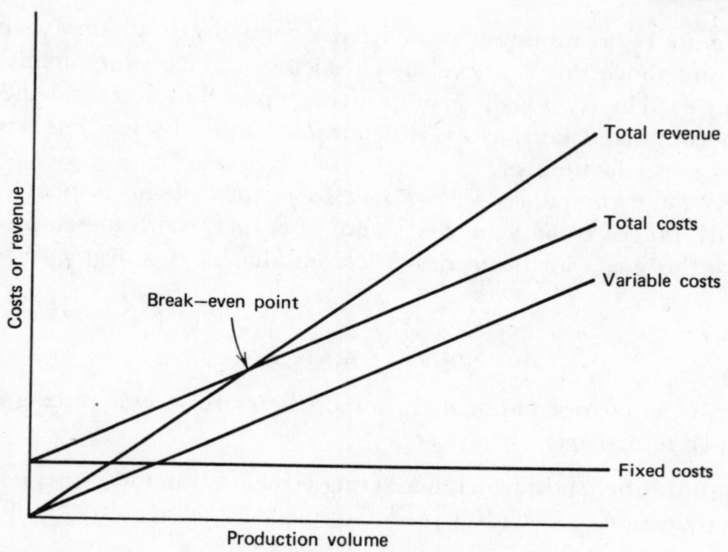

FIGURE 40. A break-even chart for product pricing (7).

units of production, the influence of the former in ensuring that the break-even point is reached and exceeded can be studied.

SHOULD THE PRICE BE MAXIMIZED?

Although one might intuitively feel that the price should be set to provide the maximum return in the short run, there are several reasons why this should not be done. High prices can attract additional competitors, who see a chance to enter the market with a lower-priced product; the interest of the government, which keep a wary eye out for antitrust activities; and the labor unions, who seize the opportunity to demand wage increases when profits are high. Stockholders would be delighted with a maximized price, but not so happy if profit levels cannot be maintained from year to year. For these and other reasons, many companies prefer to maintain the stability of roughly constant profits, even at the cost of less return than might otherwise be possible at a given time. Finally, customers realize when a price is too high, and they remember this all too well.

PROFIT GOALS

If the maximum profit is a goal which is seldom sought by responsible management, what are the targets for determining prices? Obviously, there must be a minimum profit if the company is to survive in the long run. Above this is a level of satisfactory profits, which many firms attempt to achieve. This is a subjective concept, however, and the level that is considered satisfactory is determined more by feel and attitude than as a specific number.

Many large companies set a target for return on investment, which typically ranges from 8 to 20%, and work backward to set prices to achieve this goal. Small businesses seldom aim for this degree of sophistication.

PRICING STRATEGIES

With all the factors and goals in mind, there are several pricing strategies open to management:

- Cost-plus pricing, in which marketing estimates the total costs per unit of production, and adds a predetermined percentage (say 20–40%) to set the selling price.
- Skimming the cream, in which a very high initial price is reduced in small increments over a period of time, so that the market is always satisfied at the highest possible price level. A good example is the ballpoint pen, which started at $15 in the mid-1940s but is now virtually a giveaway.
- Penetration pricing, in which the price is set low to develop as large a market as possible.
- Stay-out pricing, in which the price is set low to discourage the potential competition.
- Return on investment pricing, as described above.
- Ethical pricing, in which the company recognizes its responsibility to the public, and accepts a lower profit than it might otherwise.
- Umbrella pricing, in which firms with lower costs maintain prices high enough that even companies with higher costs can afford to stay in the business.
- Full-line pricing, in which some items in a product line are priced high so as to absorb losses incurred on other items within the line.
- Minimization of loss pricing, a desperation strategy in which the firm tries only to cover its total costs.

DISTRIBUTION

It is said (10) that manufacturing would like to make one product millions of times, that marketing would like to make millions of products once, and that the distribution people would like just enough of whatever does get made—but they have to have it yesterday.*

Distribution people are responsible for getting the product from the manufacturer to the consumer. In the American chemical industry, this is usually a direct route, with no middlemen; only a small fraction of the industry's output is sold through agents, wholesalers, or other middlemen. There are several reasons why this is practical and preferred.

First, since most of the sales of chemicals go to industrial rather than end-product markets, the relatively small number of customers can be reached with a sales force of reasonable size. And second, these customers are usually concentrated in rather small areas; one limitation on the location of potential customers of large quantities of product is shipping costs.

A third reason for direct sales is the need in the industry for extensive technical service; it would be virtually impossible for a middleman to offer the variety of necessary services described earlier in this chapter. Fourth, many chemicals are sold in very large quantities on a contract basis, in which the purchaser agrees to a fixed price for a period ranging from a few months to perhaps a year. In such cases, even the company's own salesmen may not play a part, the contract negotiations being handled by a product or sales manager who contacts a purchasing director twice a year. There is no place for a middleman here.

The distribution route must, however, be evaluated in terms of the product life cycle. In the introductory stages direct selling is essential, but at saturation the product is as well known as it ever will be, and a middleman with low distribution costs may help to stave off decline for a while by lowering sales costs. A middleman may also be used in locations where sales volume is too low to warrant the producer's maintaining a salesman, or to handle small orders, which the large producer may not find it economical to handle, lacking facilities for repackaging, billing for small accounts, and the like.

Finally, in fields other than the chemical industry the distributor performs an important financial function by paying cash to the producer, who does not have to finance a large inventory. This would be of no particular advantage to the large companies which dominate the chemical industry.

This section is based on ref. 3, while refs. 26 and 27 are also pertinent.

* And, we must add to complete the quotation, the purchasing people don't care if it even works as long as every component in it is from the low bidder.

CHALLENGES AHEAD

We close this chapter with brief mention of a few of the challenges to the American chemical industry which will pose problems for marketing organizations in the years to come. It will be no surprise to see that the same topics mentioned in several previous chapters come up here again.

ENERGY

Without doubt, all parts of the industry including marketing will have to face the challenges of the energy shortage. One can see how almost every facet of marketing will be affected—distribution, technical service, product planning, the product life cycle, pricing, and because of their close interrelations, all the rest.

FOOD (28)

"With world food production due to rise dramatically, in what nations will the rise be greatest? What will be the best locations for the plants to produce the agricultural chemicals for these nations? . . . What marketing methods will best reach the large farm of tomorrow? . . . What adjustments will be needed to meet intensified competition? . . ."

INFLATION AND PROFITS (29)

We are in a period of limited availability of capital and high costs for borrowing. There are inevitable general delays of expansions, resulting in serious shortages. Obviously, not all of the problems relate to marketing, but clear needs can be seen for improvements in market intelligence, forecasting, development, planning, and pricing strategy. Marketing management must use mature judgment in directing and controlling its operations.

GLOBAL MARKETS (8)

The trend toward multinational companies and global markets, coupled with the current economic problems of the entire world, require a thorough reexamination and rebalancing of the marketing function.

PRODUCTIVITY (30)

Competition will without doubt get keener in the years to come. We must increase productivity, and this holds for productivity of marketing personnel as well as those in R&D and manufacturing. One approach is to create an awareness of the selling profession on college campuses, in the hope that better-trained young professionals will be motivated to enter directly into marketing.

SHORTAGES (31)

In times of shortages, such as we now find ourselves in, the need for closer relations between producers and consumers becomes critical. The seller must help his customers deal with delays, substitutes, alternatives. Planning and communication in marketing must improve.

REFERENCES

1. Robert B. Stobaugh, "Chemical Marketing Research," pp. 520–526 in Herbert Popper and the Staff of *Chemical Engineering*, eds., *Modern Cost-Engineering Techniques*, McGraw-Hill, New York, 1970.
2. Irving D. Canton, "The Misunderstood Function," *Chemtech* **4**, 581–582 (1974).
3. Conrad Berenson, "Marketing in the Chemical Industry," Chap. 1 in Conrad Berenson, ed., *Administration of the Chemical Enterprise*, Wiley-Interscience, New York, 1963.
4. Arthur B. Steele, "Technical Service and Application Research—Industrial Chemicals and Plastics," Chap. 14 in ref. 5.
5. Robert L. Bateman, Symposium Chairman, *Chemical Marketing: The Challenges of the Seventies*, Advances in Chemistry Series No. 83, Robert F. Gould, series ed., American Chemical Society, Washington, D.C., 1968.
6. J. P. Cunningham, "The Challenges in Marketing for Organic Chemicals," Chap. 3 in ref. 5.
7. Conrad Berenson, *Fundamentals of Chemical Marketing*, an ACS Audio Course, American Chemical Society, Washington, D.C., 1974.
8. E. Edgar Fogle and George Forstot, "Organization for Marketing in a Large Integrated Company," Chap. 5 in ref. 5.
9. Anon., "On Sales Engineering as a Career," *Chem. Eng. Prog.* **69**, 25 (1973).
10. Gerald B. Zornow, "Some New Definitions in the Buyer–Seller Relationship," *Chemtech* **2**, 594–598 (1972).
11. Russell C. Kidder, "Marketing Research as a Guide to the Marketing Department," Chap. 17 in ref. 5.
12. Anon., "Market Research—Growth Ingredient for Chemical Companies," *Ind. Eng. Chem.* **49** (9), 42A–50A (1957); reprinted as Chap. 4 in ref. 13.
13. Conrad Berenson, ed., *The Chemical Industry: Viewpoints and Perspectives*, Wiley-Interscience, New York, 1963.
14. N. H. Giragosian, "Market Research for Planning Development and New Product Ventures," Chap. 18 in ref. 5.
15. G. T. Borcherdt, "Design of the Marketing Program for a New Product," pp. 58–

73 in R. L. Clewett, ed., *Marketing's Role in Scientific Management,* American Management Association, Chicago, 1957; reprinted as Chap. 7 in ref. 13.

16. Robert W. Schramm, "Building a Marketing Organization," *Chem. & Eng. News* **34**, 462–466 (1956); reprinted as Chap. 22 in ref. 13.
17. Herbert Popper and the Staff of *Chemical Engineering,* eds., *Modern Cost-Engineering Techniques,* McGraw-Hill, New York, 1970, Chap 8.
18. W. G. Kinsinger, "Market Development of New Industrial Chemicals and Plastics," Chap. 12 in ref. 5.
19. H. Avery and J. W. McNeil, "Market Development: Established Products, New to the Company," Chap. 13 in ref. 5.
20. George Black, *Sales Engineering, an Emerging Profession,* Gulf Publishing, Houston, Tex. 1973.
21. Monte C. Throdahl and Onnik S. Tuygil, "Sales in a World Market," Chap. 21 in ref. 5.
22. H. W. Zussman, "Application Research and Technical Service Trends in the Seventies," Chap. 15 in ref. 5.
23. Gilbert M. Miller, "Advertising Measurement—The Challenge of the Seventies," Chap. 7 in ref. 5.
24. Bernard Ganis and W. Alec Jordan, "Advertising and Publicity in the Chemical Industry," Chap. 4 in Conrad Berenson, ed., *Administration of the Chemical Enterprise,* Wiley-Interscience, New York, 1963.
25. A. D. H. Kaplan, J. B. Dirlam, and R. F. Lanzillotti, "Pricing Policy: Union Carbide," "Pricing Policy: E. I. du Pont de Nemours and Company," "Pricing Organization: Union Carbide Corporation," "Pricing Organization: E. I. du Pont de Nemours and Company," reprinted from *Pricing in Big Business,* The Brookings Institute, Washington, D.C., 1958, as Chaps. 41–44 in ref. 13.
26. Reid W. Malcolm, Jr., "Sales via Distribution and Agents," Chap. 11 in ref. 5.
27. O. E. Beutel and D. G. Griffin, "Materials Flow to the Customer—A Total System," Chap. 16 in ref. 5.
28. J. F. Bourland, "The Challenge in Marketing for Agricultural Chemicals," Chap. 1 in ref. 5.
29. W. F. Newton, "The Challenge in Marketing for Inorganic and Heavy Chemicals," Chap. 2 in ref. 5.
30. P. R. Monoghan, "The Challenge of Selling Domestic Industrial Chemicals in the Seventies," Chap. 8 in ref. 5.
31. J. R. Lee, "Marketing Techniques in a Period of Shortage," *Chem. Eng. Prog.* **70** (7), 25–30 (1974).

CHAPTER 9

STAFF DIVISIONS

Most large industrial companies, with overall corporate direction provided by the president, the board of directors, and ultimately the stockholders, are made up of nine broad functional areas which we have called divisions. In typical cases they are the research and development, manufacturing, marketing, personnel administration, corporate planning, purchasing and traffic, public relations, legal and secretarial, and finance divisions. We have already discussed the first three, the so-called operating divisions, in Chaps. 6–8, selecting them for treatment in diminishing detail in accord with the number of employment opportunities they offer to young professionals. In this chapter we turn our attention to the remaining six, which we call the staff divisions, briefly discussing the functional areas they include and available job opportunities for scientists and engineers. We have made an exception for that part of the legal and secretarial division dealing with patents, which is important enough to warrant separate treatment in Chap. 10. The general area of management is discussed in Chap. 11. Most of the descriptions in this chapter are based on ref. 1.

As shown in Fig. 2 on page 3, recent figures indicate that most chemists in industry (38%) are employed in R&D. The next largest group (34%) has advanced into management. The remaining 28% classified in the figure as "all other" breaks down into 23% in control functions and only 5% employed in other activities. Thus at best the odds are only about 1 in 20 that you obtain your first job in one of the staff divisions discussed in this chapter. Nevertheless, one or another of them may appeal to you as the place to seek an initial job, as a goal for transfer, or just in the process of getting to know your company better.

Even for these less technical divisions, the chemicals and allied products industry usually requires some technical training of its job applicants, though usually not a Ph.D. in chemistry or chemical engineering. If you are completing a B.S. course in the sciences or engineering, you might wish to consider getting an MBA degree to provide specialized training if one of the staff divisions appears especially challenging to you. Or, if you are working in R&D or manufacturing, some appropriate continuing-education courses might prove valuable before you apply for a transfer.

You may find, as many young professionals do, that your interests, outlooks, and motivations change as you grow in your company. For many reasons, an activity which appealed to you at the start may no longer hold that challenge after a while. A large company offers many other areas which may suit your growing interests better. For this reason, as well as for general knowledge should you wish to use the services offered by the staff divisions, you should be aware of their functions and the positions available in them.

PERSONNEL ADMINISTRATION

As its name implies, the personnel administration division is concerned with people. It encompasses a broad area of responsibility concerned with establishing and administering policies that affect people throughout the company. It may be called employee relations or industrial relations in some companies. Its main functions are explained below (1):

RECRUITING AND EMPLOYMENT

This division is responsible for the recruiting (including college recruiting), screening, testing, and hiring of professional and managerial personnel. (The various divisions do their own hiring of blue-collar and wageroll employees.) In a large multidivisional company, each operating unit may do its own college recruiting, but this is coordinated through personnel administration.

TRAINING AND EDUCATION

Although individual divisions usually administer orientation and training programs, personnel has overall responsibility for developing, testing, and coordinating them. It also administrates programs for scholarships, fellowships, and grants-in-aid.

WAGE AND SALARY ADMINISTRATION

This division has responsibility for developing and maintaining an equitable wage and salary structure, through job evaluation and wage and salary classifications. It prepares guidelines and procedures governing raises. The individual divisions administer local wage and salary procedures within these overall policies.

FINANCIAL BENEFITS

Personnel has responsibility for administering the so-called fringe benefit programs covering health, insurance, vacations and leaves, retirement, and the like. The finance division assists, since these programs—pensions for example—require extensive financial planning.

LABOR RELATIONS AND UNION NEGOTIATIONS

In most companies, personnel is responsible for labor relations. It is the chief advisor and coordinator in all matters involving union contract negotiation and administration and labor law. Although local plants or divisions may negotiate directly with unions, personnel has responsibility for approving contracts in advance of negotiation.

In some companies, a separate corporate division handles these functions on the theory that they require a different set of talents and lead to a different image than the company wants reflected in its personnel relations, but this practice is becoming less common.

COMMUNICATIONS

Personnel is concerned with the important area of communications through responsibility for encouraging the flow of information, up, down, and sideways. It develops oral and written communications media for use throughout the company, and may (though public relations sometimes does) prepare and publish company magazines and newspapers.

SPECIAL SERVICES

Personnel is responsible for developing and administering company programs in safety, plant security, recreation, medical services, and food services.

MANAGEMENT DEVELOPMENT

This function of personnel is really an extension of the company's training programs, designed to encourage better job performance and develop competent replacements for top-level jobs.

ORGANIZATION PLANNING

Personnel's responsibility here is in maintaining organization manuals, preparing position papers on possible reorganization, and advising divisional managers as well as company management on organization plans and changes.

BEHAVIORAL RESEARCH

If the company carries out any studies of individual and group motivation, they are likely to be the responsibility of personnel.

LEGAL RESPONSIBILITIES

The role of the personnel administration division has grown increasingly complex in recent years because of many new federal, state, and local laws affecting industry. This division is responsible for seeing that the company adheres to such regulations as those of OSHA and EEO, and to set up and administer affirmative action programs.

A good example of the organization of a personnel administration division in a large company is shown in Fig. 41.

JOB OPPORTUNITIES

Of all the corporate staff divisions discussed in this chapter, personnel administration offers the most opportunities for employment of young professionals with scientific or engineering degrees. These opportunities lie primarily in the areas of recruiting and employment, and of those special services associated with safety, health, and the environment.

Although college recruiting to fill their vacancies can be done by plant or R&D personnel, large companies may employ young scientists and engineers in corporate recruiting activities. If you can combine an ability to get along with and judge college-age people, both technically and personally, with a thorough knowledge of and honest enthusiasm for your company, this may be just the job for you.

The need for scientists trained in toxicology and industrial medicine has increased significantly in the last few years because of the spate of new EPA and OSHA regulations. Strangely, the demand for environmental engineers has not been met in the same way; for some reason, companies appear to prefer training other professionals to tackle their environment-related problems rather than hiring environmental engineers. This trend may, of course, be only temporary.

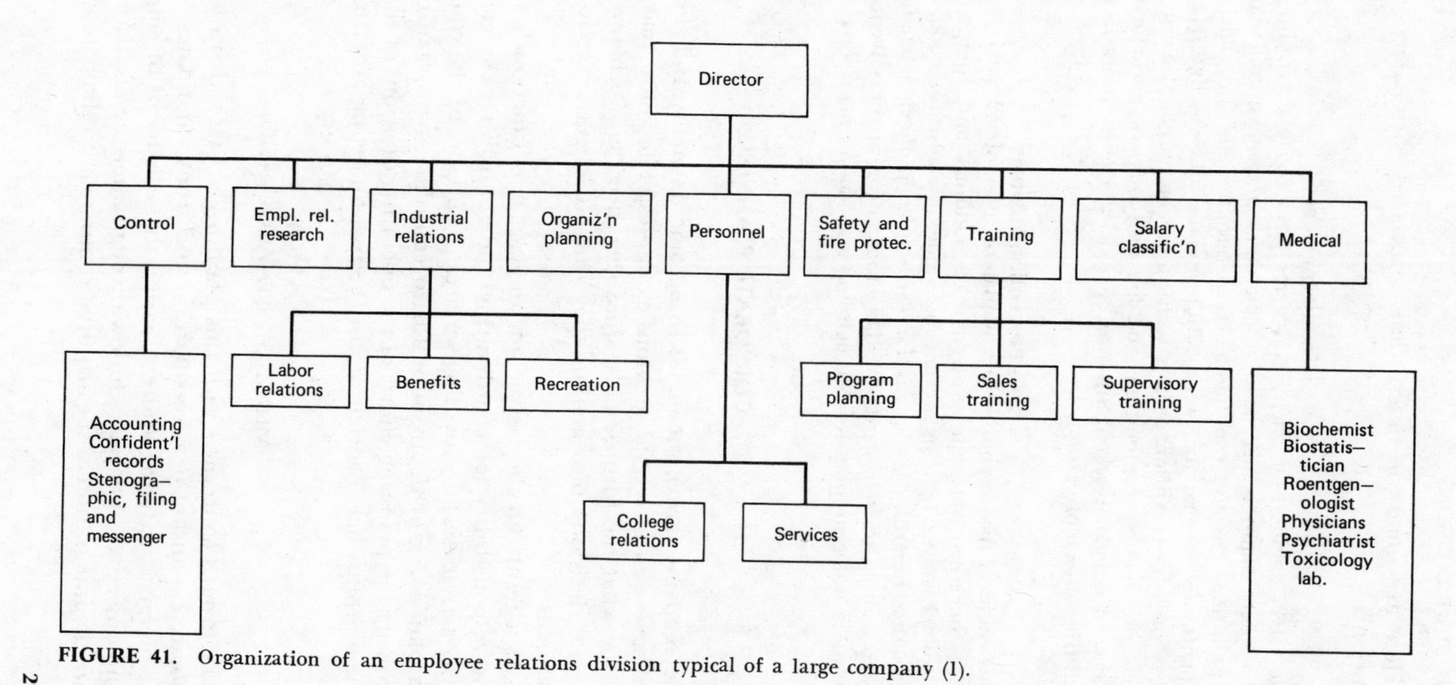

FIGURE 41. Organization of an employee relations division typical of a large company (1).

Here are some job descriptions associated with safety and loss prevention:

Safety Engineer

The safety engineer or scientist analyzes products to determine whether they meet safety requirements to an acceptable degree; environmental factors, such as pollution, are also considered. He determines the life expectancy of components, studying fatigue, wear, failure modes, and the like. He may evaluate safety or loss prevention devices such as fire or burglary alarm systems, study flame propagation, combustion, and fire resistance, or evaluate materials with respect to aging, corrosion resistance, or toxicity. He may work in R&D or manufacturing instead of personnel.

Loss Prevention Engineer

This engineer has responsibilities similar to those of the safety engineer, but is more concerned with the protection of buildings and equipment than with personnel safety. He helps prevent losses due to windstorm, flood, explosion, fire, and other hazards. He designs for earthquake protection, engineers sprinkler systems, processes safeguards for dangerous chemicals or solvents, provides for adequate water supplies for fire fighting, and studies many related areas.

CORPORATE PLANNING

As companies grow, they tend to become unwieldy and slow to react, as we have seen in several previous examples. They must make every effort to see what the future holds, since they need so much time to turn corporate decisions into action; they can no longer react to change as it happens.

As a result, in the last decade or so, a new function has evolved, taking responsibility for coordinating and controlling the company's planning activities and providing services in support of the company's efforts to plan for growth. In addition to consolidating, organizing, and controlling the planning efforts of the operating divisions of the corporation, these groups may concentrate in the two following areas:

MERGERS AND ACQUISITIONS

The groups concerned with this planning activity identify areas for corporate expansion or diversification and study candidates for acquisition or merger, analyzing and evaluating them in terms of company policies and objectives. They may follow through after the acquisition, coordinating the integration of the acquired company into the parent organization.

ECONOMICS

These units of corporate planning provide the company with analyses of general economic conditions and of the industry and markets of concern to the company, and construct forecasts of economic conditions affecting company operation.

JOB OPPORTUNITIES

The young professional scientist or engineer will find few if any positions available in corporate planning, for fairly obvious reasons.

PURCHASING AND TRAFFIC

PURCHASING

Although most purchasing is carried out by individual divisions or plants, many companies have a corporate purchasing division which provides services to the operating units. How necessary and effective this is depends on the nature of the company and the materials it uses. Corporate purchasing is most helpful in companies which are not too widely diversified, such as an oil company whose products can all be traced back to crude petroleum. In such a case the advantages of having a single purchasing unit for the entire company are obvious.

TRAFFIC

Traffic, in the sense of a corporate division, is the administration of incoming and outgoing shipments. It is easy to see why purchasing and traffic—which of course may exist as separate divisions—are so often consolidated.

The corporate traffic unit provides advice and service to the operating units and the company as a whole in the following ways:

- Ensuring an adequate supply of truck trailers, freight cars, barges, or other transportation when and where it is needed.
- Negotiating with freight carriers with respect to rates and contracts, for both incoming raw materials and outgoing products.
- Producing traffic manuals for use by the operating units.
- Administering claims against shippers for loss, damaged goods, or errors in billing.
- Representing the company in government or other judicial matters involving traffic.
- Administering the company's fleet of automobiles, planes, or boats.

JOB OPPORTUNITIES

Both chemists and chemical engineers can occasionally find jobs in this division, as purchasing agents for example, but the opportunities are limited.

PUBLIC RELATIONS

The role of public relations is viewed quite differently in various companies. Some look upon it as a means of furthering the objectives of a specific phase of their activities, and assign it to the corresponding division. Examples include marketing, personnel, and finance. In many other companies, however, the emerging importance of the corporation's role as a good citizen of the community has led to the establishment of a public relations division. The size of such a division varies widely, from a mere handful in a large chemical company which, however, produces only basic chemicals, sold to other companies, to the vast networks of, say, the big three automobile producers.

In the latter cases one finds three types of subdivision of the public relations division; often all three are practiced within the same division. One type of unit groups people with certain skills, such as motion picture or publications; another is concerned with a certain portion of the public, such as stockholders, educational institutions, or dealers; and the third consists of geographic or regional groupings based on local needs.

A public relations division usually has four major areas of responsibility:

- It builds and maintains good relations with those parts of the public whose attitudes are of concern to the company.
- It prepares top management speeches, special exhibits, and publicity programs and events.
- It produces publicity releases, announcements, and publications that go outside the company, in some cases sharing responsibility for both these and internal publications with personnel administration.
- It maintains contacts with the communications media—newspapers, radio and television, magazines, and motion pictures.

In addition, to the extent that the company benefits from advertising itself instead of its products, some advertising functions may be assumed by public relations—after all, the skills and contacts required for the two functions are similar. Public relations may also include a group responsible for representing the company in civic and governmental affairs.

JOB OPPORTUNITIES

If you are blessed with the ability to communicate with unusual skill, both orally and in writing, don't overlook the possibilities of a job in public relations. There is some need for both scientists and engineers.

LEGAL AND SECRETARIAL

Although the by-laws of virtually every organization call for a secretary, not all large companies have secretarial divisions, nor do all have legal divisions. But when one is missing, its functions are usually carried out by the other, and the two are often combined, as indicated in this section.

THE SECRETARIAL FUNCTION

The typical responsibilities of a corporate secretary fall into three areas:

- Services to and for the board of directors, such as keeping records and minutes, and issuing calls for meetings.
- Keeping records of stock certificates and other corporate documents, and giving final legal approval to contracts, stock, deeds, and other documents by affixing the corporate seal to them.
- Maintaining responsibility for stockholder relations and correspondence, and issuing the annual report and similar documents.

In addition, the secretary's group often coordinates the activities of such other staff units as office services, public relations, and personnel administration, and in a good many companies the secretary is also treasurer.

As the secretary of a very small professional society, one of us can state that these formal responsibilities represent only the tip of the iceberg, underlaid by a continual flow of duties, large and small, important to the well-being of the organization.

THE LEGAL FUNCTION

The purpose of the corporate legal unit is to ensure that the objectives, policies, programs, and actions of the company are at all times fully within the confines of the law—or in less positive terms, as stated by one corporate executive, "to keep the company out of jail." Some of the areas in which legal services to the company are important are patents, contracts, government relations, litigation, real estate, labor relations, antitrust, and taxes. To the young professional, patents are without doubt the most immediate and important of these, and we recognize this by devoting Chap. 10 to them.

JOB OPPORTUNITIES

As mentioned further in Chap. 10, many chemists and engineers find a particular challenge in patent and related work and obtain, often in continuing education, a law degree allowing them to enter this phase of industrial employment.

FINANCE

The one staff division present in every company, usually ranked next in importance to the president, is finance. Its duties include:

- Establishment of financial polices.
- Assignment of financial responsibility.
- Development of standard procedures for accounting, analysis, and reporting financial data.
- Controlling and approving all major actions and important changes in procedures related to finance.

In addition, the financial staff provides services and guidance in financial matters to the operating units. These responsibilities are in most cases divided between two branches of the finance division: the treasury and controller functions.

THE TREASURY FUNCTION

This group, headed by the traditional treasurer, concentrates on the provision of an adequate supply of operating funds. It is concerned with cash management (floating corporate stock and bond issues, maintaining a market for company securities, securing adequate supplies of short-term capital), credit management, bank relations, insurance, and employee benefits.

THE CONTROLLER'S FUNCTION

The position of controller has grown out of the financial control provided by accounting, and accounting (which is described for the chemical industry in ref. 3) remains the most important responsibility of the controller's group. Over the years, however, it has added responsibility for budgetary planning and control, internal auditing, systems procedures for financial information and reporting, and taxes.

JOB OPPORTUNITIES

There is little or no need for professionally trained scientists or engineers in the finance division.

REFERENCES

1. Harold Stieglitz and Allen R. Janger, *Top Management Organization in Divisionalized Companies,* Personnel Policy Study No. 195, National Industrial Conference Board, New York, 1965.
2. Anon., "Job Outlook Tight for '73," *Chem. & Eng. News* **50** (40), 10 (Oct. 2, 1972).
3. Jack W. Mueller, "Accounting and the Chemical Business," Chap. 8 in Conrad Berenson, ed., *Administration of the Chemical Enterprise,* Wiley-Interscience, New York, 1963.

CHAPTER 10

PATENTS

Since patents have been mentioned frequently in earlier chapters, we assume that you are now familiar with the fact that one of your major goals as a scientist or engineer in industry is to apply your knowledge creatively to promote new technology by generating new products, processes, and compositions, and that this new technology is most often protected through patents.

Patents are of great importance to both you and your company. They serve a major function in both stimulating and protecting the company's investment in research and development and assuring its continuing introduction of new and improved products in the marketplace. For you, patents are tangible evidence of creativity, and patents in your name can significantly affect your advancement.

Of all the U.S. patents granted in the last decade, over 20% dealt with chemical inventions, and, although exact figures aren't available, it's clear that the vast majority of these patents were granted to persons or groups of persons working in industry. With almost 4 million U.S. patents issued by 1975, it follows that the patent literature is an excellent source—and probably the largest single source—of new technology, the latest developments, and the stimulation of new ideas.

Despite this evidence of their importance, it is both surprising and in some ways shameful that patents and patent law are almost totally neglected as subjects for study in most chemistry and chemical engineering curricula. We feel that a basic understanding of the U.S. patent system is highly desirable for all technically trained people. In this chapter we can only skim the highlights contributing to such an understanding.

THE PATENT SYSTEM

What more appropriate beginning of any discussion could there be than to quote the origin of the subject in the U.S. Constitution? Article 1,

Section 8, says that "Congress shall have power. . . . To promote the progress of science and useful arts, by securing for limited time to authors and inventors the exclusive right to their respective writings and discoveries." With respect to authors and their writings, the mechanism for securing these rights is the copyright law, described later in this chapter; for inventors, the rights are secured through patents granted by the Patent Office of the Department of Commerce. Once a patent has been issued, the inventor (or co-inventors), or his (or their) assignee, has the legal right to exclude others from practicing the invention (defined by a claim or claims of the patent) for a period of 17 years from its date of issue, to license others to practice it, or to sell the rights to someone else.

These proprietary rights serve a very useful purpose as an incentive for invention. Publication of the new technology via the patent, which is printed by the Patent Office and sold for 50¢ a copy, benefits society by stimulating new ideas which, in turn, may lead to new inventions. Without patent protection for inventions, most of the incentive for the free-enterprise system would vanish: Wherever possible, new technology would be retained as trade secrets, and no new product would be safe from immediate copying.

In the United States, patents are granted on applications filed in the name of the individual inventor or inventors. The inventor(s) may do a variety of things with the patent application or patent, as indicated above, including assigning (i.e., selling) or licensing it. As discussed in Chap. 4, most companies protect their considerable investments leading up to the patent by requiring each professional employee to sign an employment agreement in which the employee agrees to assign all (job-related) patent applications and patents to the company.

While a patent grants the inventor the right to exclude others from making, using, or selling his invention, it does not give him the right to practice the invention, contradictory as this may seem. Whether or not he can do so depends on whether the invention, as practiced, would infringe on the rights of another inventor or his assignees under any other patent. Let's cite an example:

Suppose that you, working for Company A, discover that a certain new phosphate surfactant will impart improved stability to acrylic latex paints. Your company obtains a patent naming you as the inventor, and you assign your rights to the company. Company A now has the right to prevent others from using this particular surfactant in acrylic latex paints. Later, an inventor in Company B discovers that, when formulated with a specific silicone surfactant, the new phosphate gives not only improved stability but better spreading characteristics. Company B obtains the ownership of a patent on this combination. But Company B cannot practice its invention without obtaining a license from Company A. And conversely, Company A cannot use the specific combination of the two surfactants without a license from Company B.

A patent issued by the Commissioner of Patents is presumed to be valid. However, just because a patent is granted does not necessarily mean that it may not later be found invalid during litigation in a court. It may be that the Patent Examiner in the Patent Office did not consider all the prior art, because all of it may not have been located by him, or the inventor may have publicly used his invention or sold it more than one year before filing his application for a patent. The major cause of the invalidity of patents is said (1) to be the lack of a patentable invention as defined by the patent laws (and discussed below). A lawsuit is often brought with the objective of proving a patent invalid; this is becoming more and more common as a defense against claims of infringement. Such suits are almost always long and very, very expensive.

Of course, the fact that a patent has been granted by the Patent Office says nothing about the value or marketability of the invention. Many patents are worth very little, and others have but modest commercial significance, but the few worth millions of dollars receive considerable publicity and often create new industries. However, it should be kept in mind that a patent can only be as good as the invention underlying it and, even then, its financial value will depend in part on economic and social conditions. For example, a basic patent on television might be valueless if the public did not have ready access to electric current.

Finally, by giving the inventor exclusive control over his invention for 17 years, a patent creates a private property right of limited duration, which can be put to use to benefit society as well as to make a profit for its owner. It is here that industry finds a major incentive to risk large amounts of money in an effort to innovate new products and processes, and to risk additional substantial amounts to establish manufacturing and marketing facilities for such innovations.

THE COMPANY PATENT DEPARTMENT

From the time an invention is conceived to the time a patent is granted by the Commissioner of Patents in the U.S. Patent Office in Washington, the burden of activity and proof is on the inventor and his company. Most large companies have a patent department or division to direct this activity. Often, it is a branch of the legal department, with a high-ranking company official as its head.

The main responsibility of the patent department is to ensure that the operations of the company are properly conducted from the patent standpoint. Through its patent attorneys, the patent department conducts activities such as applying for and obtaining patents, examining new developments, seeking or granting licenses, handling patent litigation problems (generally through outside legal counsel), and assessing company operations for possible infringement of existing patents of others. It also has the important responsibility of supplying information on patents and related areas, often referred to as the "state of the art," to researchers, the professional staff, and management.

In a large company with such a patent department, the procedures for reporting and assigning inventions and documenting discoveries are well established, and are made known to new employees soon after their employment. The patent department is there to be called on at all times —and you should not hesitate to take advantage of this. Even the need for interpreting the meaning or breadth of claims in an existing patent which may be related directly to your work is more than adequate reason to call on the legal experts in the patent department for advice, without delay. Don't try to do this yourself, and don't write your own interpretation of the scope or validity of patent!

Small companies, lacking the resources to support a full-time legal and patent staff, may utilize the services of independent patent attorneys or law firms, either on a permanent basis or as the occasion arises. Such services are also valuable to large companies to supplement their own staffs, as required.

It's worth noting that many chemists and chemical engineers find patent law an interesting, challenging, and thoroughly satisfying career. Often they go to law school, either full time or at night during industrial employment, after obtaining a B.S. or M.S. degree in their scientific discipline. Many companies encourage and support this route in order to develop the technically trained patent attorneys their patent departments desire.

STEPS IN OBTAINING A PATENT

In this section we wish to discuss the important steps of invention and procedures involved in obtaining a U.S. patent. They are outlined in Fig. 42 to provide an easy reference for the discussion.

WHAT IS A PATENTABLE INVENTION?

Invention begins with an idea, but obviously not every idea leads to a patentable invention. The conditions for an invention to be patentable are defined in the U.S. patent laws, specifically Sec. 100-104 of the U.S.

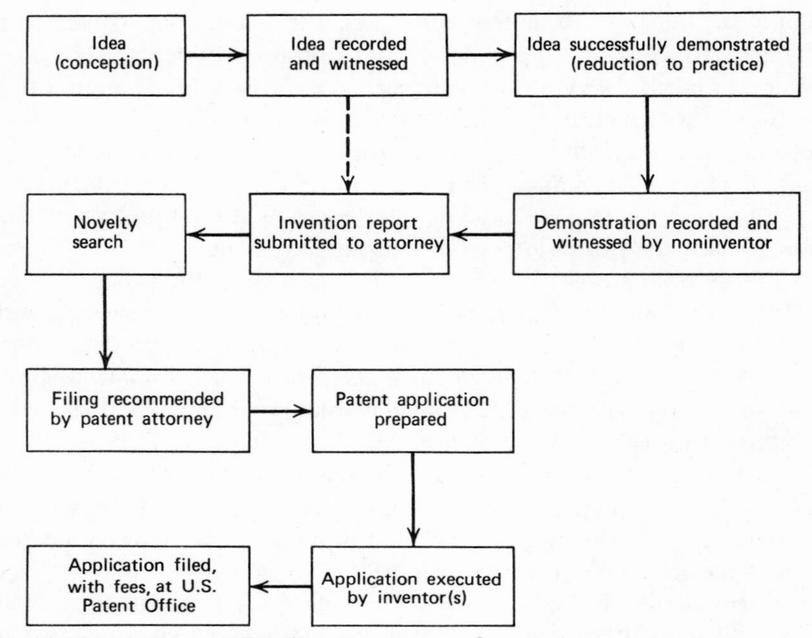

FIGURE 42. Important steps in the invention-to-U.S. patent process.

Code, Title 35, Patents (also referred to as the Patent Act of 1952) (2). Section 101 defines patentable subject matter. It states that "whoever invents or discovers any new and useful process, machine, manufacture, or composition of matter, or any new and useful improvement thereof, may obtain a patent therefor, subject to the conditions and requirements of this title." Note the four classes of subject matter spelled out: process, machine, manufacture, and composition of matter. Note also that the invention may be "invented or discovered," and must be "new and useful" (Fig. 43).

Section 103 adds the further qualification that the invention must *not* have been *obvious*: "A patent may not be obtained . . . if the difference between the subject matter to be patented and the prior art are such that the subject matter as a whole would have been obvious *at the time* the invention was made to a person having *ordinary skill* in the art to which said subject matter pertains. Patentability shall not be negatived by the manner in which the invention was made."

Novelty, timing considerations, and the right to patent are spelled out in Sec. 102, which states that a person shall not be entitled to a patent if:

(a) the invention was known or used by others in this country, or patented

FIGURE 43. From *Ind. Res.* **16**(11), 33 (November 1974). Reprinted by permission of *Industrial Research* magazine.

or described in a printed publication in this or a foreign country, before the invention thereof by the applicant for patent, or

(b) the invention was patented or described in a printed publication in this or a foreign country or in public use or on sale in this country, more than one year prior to the date of the application for patent in the United States, or

(c) he has abandoned the invention, or

(d) the invention was first patented or caused to be patented by the applicant or his legal representatives or assigns in a foreign country prior to the date of the application for patent in this country on an application filed more than twelve months before the filing of the application in the United States, or

(e) the invention was described in a patent granted on an application for patent by another filed in the United States before the invention thereof by the applicant for patent, or

(f) he did not himself invent the subject matter sought to be patented, or

(g) before the applicant's invention thereof the invention was made in this country by another who had not abandoned, suppressed, or concealed it. In determining the priority of invention there shall be considered not only the respective dates of conception and reduction to practice of the invention, but also the reasonable diligence of one who was first to conceive and last to reduce to practice, from a time prior to conception by the other.

Some of these points benefit from restatement in clarified form. A provision of great practical importance in U.S. patent law is that a patent *cannot be obtained* if the application is filed more than *one year* after the invention has been described in a printed publication here or abroad or put on sale or used publicly in the United States. Truly tragic losses of patent rights have occurred by letting the one-year "statutory bar" period for filing a patent application pass through inadvertence or oversight. The law on this point differs markedly from country to country.

It is clearly indicated that only the inventor can obtain a patent. Thus it's important to determine accurately who the inventor is before submitting the patent application. We shall consider this problem later.

The subject matter on which patents can be obtained is described in Sec. 101 as "any new and useful process, machine, manufacture, or composition of matter," and it is best to define these terms as they are used:

A *process* is a method of producing a physical result through one or more operational steps. It may also include a new use of an old material.

A *machine* is a device or combination of devices by which some operation is performed.

The term *manufacture* refers to a man-made product other than a machine or composition of matter. The discovery of a naturally occurring product is not patentable, although various aspects of modifying, treating, or using the product may be patentable.

A *composition of matter* is a chemical composition, including a compound or a mechanical mixture of two or more ingredients.

In the United States, an invention must be demonstrated or "reduced to practice" in order to be patentable, though models are not necessary. This is not a serious restriction, however, since even untested ideas can be patented because the written patent application is considered legally as "a constructive reduction to practice." That is, it is legally accepted as a disclosure of the invention for obtaining a patent. But an inventor who does not demonstrate his invention experimentally may overlook factors which could affect the breadth of his claims and the utility of the patent.

Finally, a patent will not be granted if the invention, once made, has been abandoned by the inventor. If a patent is later contested, and the time between conception and filing the patent application is too long, the court may rule that the invention was abandoned.

RECORDS AND HOW TO KEEP THEM

Proper records can be vital in obtaining a patent. Perhaps their most important use is in determining the actual date of an invention and in settling matters of priority between different inventors in interferences between patent applications. The actual date of the invention is the date on which the inventor first reduced the invention to practice, or filed his patent application, but the law provides that this date may be related back to the date of conception *if diligence* (which must be clearly shown) is exercised in reducing the invention to practice and in filing. The best proof of the date of conception is written documentation, signed and dated by the inventor and suitably witnessed by a corroborating witness who actually saw the inventor make or practice his invention on that date. The importance of proper record keeping in any work which may lead to a patent is thus obvious.

Accurate written records of research and development work should be kept for a number of reasons in addition to documenting the date of an invention. Among them are:

- To minimize costly duplication of effort.
- To aid in patent prosecution.
- To provide proof of the validity of a patent, should it be required by the courts.

The best place to record ideas and information is in a bound laboratory research notebook whose pages are numbered consecutively. Such notebooks are available from many suppliers and are stocked by most large stationery stores. Most companies large enough to have R&D and patent departments provide notebooks with the company name on each page and instructions for using the notebook in the front. Table 24 shows an example, listing most of the things that should be kept in mind in recording information.

One of the most important points about record keeping is the requirement that records be properly signed, dated, and witnessed: Unwitnessed notebook records, and often the inventor's oral testimony, are usually insufficient evidence in court. The corroboration of the inventor's record of his idea is important for establishing dates of conception and actual reduction to practice. Someone who is not a co-inventor should observe what the inventor has done and read and fully understand each research notebook entry, then promptly sign and date it as a witness. Ideally and if practical, a noninventor should repeat the experiment and properly record his results in a research notebook. If others have contributed to the idea, their contributions should be acknowledged in the record.

Sometimes information pertinent to the invention is sent or received separately, as by correspondence or reports from service groups. This material can be entered permanently into the notebook by taping or pasting it in place (with the notice "nothing underneath" written along-

TABLE 24

Typical Research Notebook Instructions

The XYZ Company
Laboratory Research Notebook
Notebook No._______

Instructions for Using Notebook

1. Entries must be in ink, and should be neat and easily legible.
2. The pages of the notebook should be used in consecutive order so that the dates on the pages run chronologically.
3. Entries should be made in the notebook at the time the work is performed.
4. Avoid erasures. If a mistake is made, cross out and enter the correct fact.
5. Each page of the notebook should be dated (month, day, and year) and signed at the bottom after the printed word "Signature" by the one doing the work. The notebook entry and the signature should be made on the same day the work was done. Each page should be read, signed, and dated at the bottom by a witness, preferably on the day on which the entry was made.
6. If it becomes necessary to make an entry at a later date on a notebook page which has already been dated and signed, the subsequent entry should be dated and duly signed and witnessed.
7. Record all steps in sufficient detail so that any person skilled in the art to which the record pertains can repeat your work and obtain the indicated results.
8. Record the materials used, the chemical composition if known, quantities and concentrations (in standard chemical terms), solvents, temperatures, times, the order of mixing, significant pressures, etc.
9. Only standard abbreviations should be used, that can be found in an authoritative dictionary or in standard lists of abbreviations, such as those

side), or alternatively kept in a separate looseleaf notebook with adequate references in the research notebook.

In marketing and other less technical areas, ideas are usually reported in correspondence related to specific products or markets. Reports like this should be filed, and the files periodically microfilmed to provide a permanent and less bulky record. Ideas which might lead to patents should, of course, be signed, dated, and witnessed at the time of conception.

In writing up the research record, time should be taken to include analysis, conclusions, and summaries of major points and their implications. All the information should be recorded initially in the research notebook, *never* on separate sheets of paper for later transcription.

published by the American Chemical Society or by the American Institute of Physics.

10. Code terms or numbers may be used with one or the other of the following provisions:
    (a) That a master book containing a glossary is permanently available to the inventor and available to the witnesses to his records. Both inventors and witnesses must understand the full meaning of all code terms and numbers used in the record. The master code book should be kept under the jurisdiction of a particular individual who should sign and certify the date when the book was set up, and initial and date each subsequent entry.
    (b) That an explanation of the term or number is given in standard chemical terms at least once on the page where the term or number is used.
11. For data that can not be recorded conveniently in this bound notebook, pads of separate work sheets printed like those in the notebook are available. These may also be used for making carbon copies of entries in this standard notebook. The separate sheets are furnished to fit standard three-hole notebook covers, or they may be filed in manila folders, etc. Other supplementary records, graph sheets, etc., should
    (a) be identified as Company property, for example, by the words XYZ Company, R & D Division,
    (b) be signed and dated by the worker and a witness, and
    (c) contain a specific reference to the notebook number and page containing the related record.

Signature____

From____________________ to____________________

Date Date

THE INVENTION REPORT

After a presumably patentable invention has been reduced to practice and its utility demonstrated, a formal invention report is usually written to the patent department by the inventor, containing a detailed disclosure of the invention. Most large companies have standardized forms and procedures for preparing and submitting invention reports. This report should provide background information which will help a patent attorney determine whether or not the invention is new, useful, and unobvious. It should state how and when the invention has been reduced to practice. It is also desirable that it provide an assessment of the possible value of the invention to the company. Of course, the report will be based on written records duly signed, dated, and witnessed in research notebooks.

It is the responsibility of the inventor, being "skilled in the art," to identify and provide all of the prior art of which he is aware which he believes is pertinent to the invention and could affect its patentability. This means that he should carry out a search of the published literature —journals, patents, textbook, and all other sources—with the assistance of the company's information services groups. It's likely that most of this study has already been made in the course of the research leading to the invention, so that the task of the prior art search is not as burdensome as it might seem.

Since the invention report must be written by, or at least unequivocally identify, the inventor(s), now is the time for all those concerned to get together to review their contributions, define the invention precisely, and make a preliminary decision as to who the actual inventors are. This can be a difficult task which should be carefully reviewed with patent counsel, and we discuss it further in the next section.

The invention report typically includes the following:

- Title and a brief abstract.
- The inventors' names, places of employment, and citizenship.
- The field to which the invention pertains.
- A concise description of the most relevant prior art known to the inventor.
- A concise statement of the invention, its objectives and its advantages, including, if known, prior attempts to solve the same problem and the reasons why those prior attempts failed or were imperfectly successful.
- Detailed examples of the practice of the invention, supported by illustrations and diagrams as required, and their locations by notebook numbers and pages.
- A discussion of any variations that might be employed in carrying out this practice—this is of considerable importance if patent protection of adequate breadth is to be obtained.
- A review of the history of the invention, documenting the timing, people, and records involved in completing the invention from its conception.

The distribution of the invention report is usually very restricted, since it contains confidential company information which, if accidentally disclosed, could jeopardize the patentability of the invention and much of its value to the company as either a trade secret or a patent.

Submitting an invention report does not guarantee that it is worthwhile for the company to seek patent protection for the invention. Several other questions must be considered by the patent attorneys and by management to decide if patent protection is warranted, such as:

- Is the invention likely to be utilized by the company?
- Does it have potential value for licensing to others?
- Is the value of the new technology great enough to warrant the cost of obtaining a patent?
- Should the invention be patented in other industrial countries?
- Would the company's interests be better protected by maintaining the technology as a trade secret, or by disclosure in a technical publication for the benefit of society or to keep the way clear for the company to practice the invention in the event a later inventor makes the same invention and tries to patent it?

SELECTING THE INVENTORS

In the United States, patents are granted only to the first and original inventor(s), and the patent application must be made in the inventor(s) name(s). Today, it is rare that an invention is conceived and reduced to practice by a single person. It is much more likely that several people will have contributed to the successful culmination of a project, and it is often an involved process to determine the identities of the actual invent-or(s). The cardinal rule is that he who does the thinking and conceiving is the inventor, not he who simply carries out instructions, even though the latter expends considerable time and effort in doing so. Saunders (3) gives several examples of the complex and confusing situations that may arise.

In analyzing the roles of all those who have contributed technical assistance to the project, those who have made direct contributions to the patentable aspects of the invention must be identified. They become the co-inventors, and all others do not. However, the latter may become good corroborating witnesses. The patentable aspects must be identified in detail and must be clear to all involved if misunderstandings and hard feelings are to be avoided. As noted above, even if someone carried out your ideas, you are the sole inventor if no new invention is required in his doing so. If he expands on your ideas while carrying them out, making his own contributions to the patentable features, then there may be a joint invention or there may be two individual inventions. In any case, the patent department will help to make the decision as to who should be named as inventor(s) on the patent application if one is filed.

COMPENSATION TO THE INVENTOR

As we've said before several times, most companies require their professional employees to assign to the company the rights to all patents obtained as a result of the employee's job responsibilities. Most employees agree with this policy, feeling that technical innovation is a major part of a professional's job responsibility; because of this, the cost of obtaining patents, and the cost of providing employees with a creative work environment, the rights to patents developed on the job properly belong to the company. (On the other hand, some companies claim full rights to all patents obtained by their employees, even those on inventions not associated with the job, made independently by employees on their own time and at their own expense. Many professional employees feel, however, that the company should not have exclusive rights in these instances.)

In Chap. 5 we mentioned the various modes of direct and indirect compensation for patents from which employees are likely to benefit, including bonuses and more rapid advancement. Despite these benefits, there are cases where an inventor feels that his compensation has been inadequate in view of the value and significance of the invention, and that the patent assignment agreement he was required to sign is too restrictive. This is a complex problem (4), and from time to time federal legislation has been proposed (5) to change patent assignment practices.

Patent compensation practices were reviewed by an ACS *ad hoc* Committee on Economic Status (6), using a survey of 140 industrial employees. The committee found that while many companies provide monetary recognition to inventors on both filing and issuance of the patent, the most widely used method of compensation was indirect, via salary raises and promotions. They concluded that "companies arrived at their practices not unmindful of employee needs and feelings..." and recommended "that all employment agreements should limit the assignment of patent rights to areas related to the company's present and anticipated business activities."

THE PATENT APPLICATION

Upon receipt of your invention report, your patent attorney will become familiar with the technical details of your invention. He will review and extend your prior art search, paying careful attention to previous patents. He will then prepare an analysis for final management approval, including the claims (patentable features) which he thinks the patent office will allow. If approval is received, he then prepares the patent application to be filed with the U.S. Patent Office.

The patent application has two major parts and several attachments. The first major part is the *specification,* a description of the invention and the prior art leading up to it, including examples and illustrations; the second consists of the *claims,* brief explicit descriptions of the new and essential features of the invention. The application must be accompanied by an *oath,* the inventor's sworn statement that to the best of his knowledge he is the original and first inventor, or a *declaration,* an unsworn statement to the same effect. The *filing fee* of $65 plus additional charges based on the number of claims is paid at this time also.

The details of writing up the patent application following specific formats and procedures will be handled by the patent attorney, but the inventor will review and approve the final draft for technical content before signing it. This review and approval should never be considered a routine chore, and in fact it is essential that good communication and cooperation be maintained between the inventor and the patent attorney from the time the invention report is written until the patent is granted.

The information required in the patent application for chemical patents (1) is similar to that in the invention report, discussed earlier. The detailed information describing the specification and claims is summarized (7) by the Patent Office as follows:

> The specification must include a written description of the invention and of the manner and process of making and using it, and is required to be in such full, clear, concise, and exact terms as to enable any person skilled in the art to which the invention pertains, or with which it is most nearly connected, to make and use the same.
>
> The specification must set forth the precise invention for which a patent is solicited, in such manner as to distinguish it from other inventions and from what is old. It must describe completely a specific embodiment of the process, machine, manufacture, composition of matter or improvements invented, and must explain the mode of operation or principle whenever applicable. The best mode contemplated by the inventor of carrying out his invention must be set forth.
>
> The claims are brief descriptions of the subject matter of the invention, eliminating unnecessary details and reciting all essential features necessary to distinguish the invention from what is old. The claims are the operative part of the patent. Novelty and patentability are judged by the claims, and, when a patent is granted, questions of infringement are judged by the courts on the basis of the claims.

The Patent Office booklet (7) also includes copies of the several forms to be signed by the inventor, among them being the application, the oath or declaration, a power of attorney, and the assignment of the patent.

PATENT OFFICE PROCEDURES

When the patent application is formally filed by submitting it to the Commissioner of Patents in Washington, the process of "patent prosecution" begins. Its several stages, which can be both lengthy and complicated, are outlined in Fig. 44. They are described more fully in a recent article (8), and we will only say that there is usually a good bit of action between the Patent Examiners in the Patent Office and the attorneys representing the inventor before a fully acceptable set of claims is arrived at or the patent is finally rejected in its entirety. During all this time, which can take several years, the application is kept in strictest secrecy by the U.S. Patent Office.

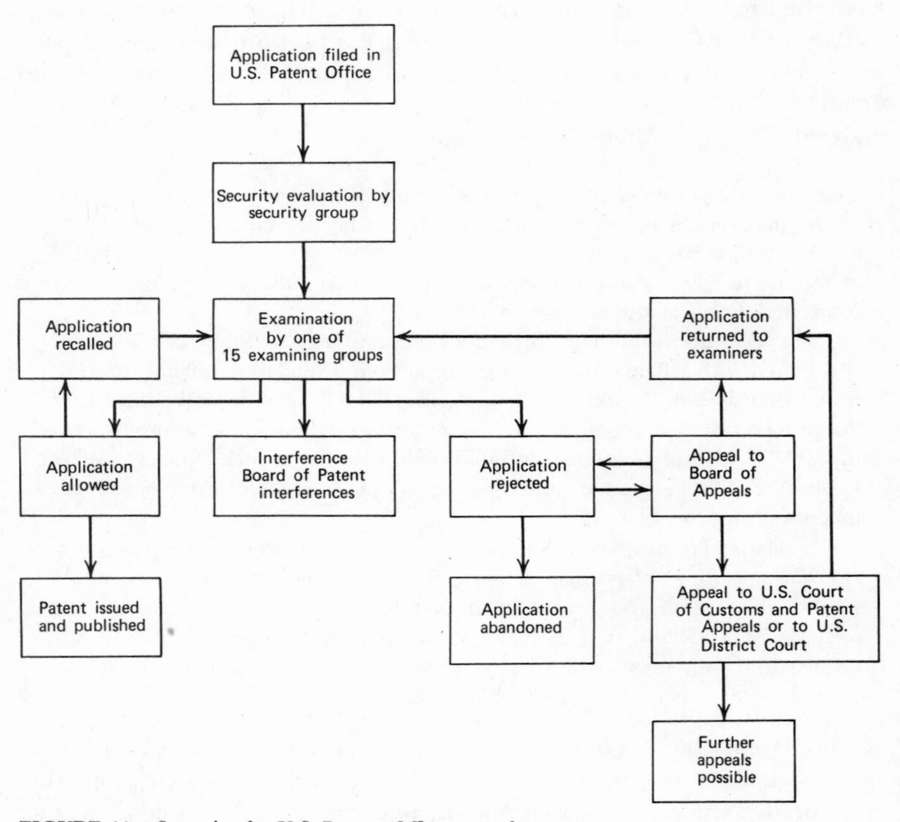

FIGURE 44. Steps in the U.S. Patent Office procedures.

Patents may also be obtained under the laws of other countries. In most cases it is beneficial to file in such countries within one year after filing the U.S. application. It is not unusual for one or more such foregin applications to be published before the U.S. application is published. The patent laws of each country are different, hence patent matters outside the United States are often handled by an attorney who specializes in such work. An interesting system now being devised to unify patent laws in the European countries is described in ref. 9.

THE PATENT LITERATURE AND HOW TO USE IT

In addition to becoming familiar with the basic information about patents outlined in this chapter, we feel that professionals entering industry should know (a) where to get more details when they are needed, and (b) how to use the patent literature.

For general information on patents, we suggest reading one or more of the excellent and inexpensive booklets published by the Patent Office (2, 7, 10, 11). You may also wish to refer to recent books on the subject (12, 13).

The ACS book *Patents for Chemical Inventions* (14) is a good start in seeking information on patents and the chemical industry. Encyclopedia articles and chapters in other books are also recommended (1, 3, 15).

The Kirk–Othmer Encyclopedia of Chemical Technology contains an excellent review of the U.S. and international patent literature (16), including a discussion on how to keep up-to-date. Perhaps the best way to do this, however, is to read the *Official Gazette* of the U.S. Patent Office (17). This weekly periodical publishes the abstracts or claims of all U.S. patents issued during a given week; most large libraries subscribe to it. *Chemical Abstracts* is another excellent source for keeping abreast of the chemical patent literature. Finally, copies of individual patents can be purchased for 50¢ from the Commissioner of Patents, Washington, D.C. 20231.

TRADEMARKS AND COPYRIGHTS

In addition to but quite separate and distinct from patents, the U.S. Government offers two other forms of protection against the copying or misuse of an individual's or a company's rights or property. These are trademarks and copyrights.

TRADEMARKS (15, 18, 19)

The Trademark Act of 1946 (Chap. 22, Title 15 of the U.S. Code) defines a trademark in Sec. 15 as including "any word, name, symbol, or device, or any combination thereof adopted and used by a manufacturer or merchant to identify his goods and distinguish them from those manufactured or used by others." The Patent Office booklet on Trademarks (18) goes on to point out that, although their primary function is to indicate origin, trademarks "also serve to guarantee the quality of the goods bearing the mark and, through advertising, serve to create and maintain a demand for the product." This description certainly seems to fit such well-known trademarks as Coke, Orlon, Prestone, Xerox, Kodak, or Seagram's.

The fact that trademarks are registered by the Patent Office, sometimes indicated by the phrase "Reg. U.S. Patent Office" or by the symbol ® beside the mark, should lead to no confusion between trademarks and patents; they are entirely different, of course.

The rights to a trademark are acquired by its use in commerce, and this use must continue if the rights are to be preserved. Furthermore, the owner of a trademark must maintain constant control over its use, and admonish others against its misuse, or his rights may be lost. Aspirin, escalator, and cellophane are words which were once valuable trademarks but whose rights in the United States have been lost to their original owners.

Unlike a patent, which has a life of seventeen years and is not renewable, a trademark registration issued under the Act of 1946 remains in force for twenty years, and may be renewed for a further period of twenty years unless previously cancelled or surrendered.

COPYRIGHTS (15, 20)

Like a patent in that it finds its origin in the Constitution, a copyright grants to an author the exclusive right to produce, publish, and sell multiple copies of his writing, including literary, dramatic, musical, and artistic as well as scientific works. Copyright protection is obtained simply by publishing the work with a notice of copyright properly located on it. In the case of a literary work, this should be on the title page or the page immediately following. A suitable notice uses the word "Copyright" or the symbol ©, and might for example read

Copyright © 1975 by John Wiley & Sons, Inc.

Failure to include such a notice on the first publication and all thereafter will, as a general proposition, preclude any possibility of obtaining copyright protection.

It is customary for the author, or more often his publisher, to register the copyright for an initial term of 28 years, by sending two copies of the work, together with an application form and a small fee, to the Register of Copyrights at the Library of Congress in Washington. The Patent Office has nothing to do with copyrights.

There is growing concern in the publishing business over the increasingly widespread use of office copying machines to make multiple copies of copyrighted matters, such as scientific material or journal articles (21). The difficulty is that to date no one has figured out how to enforce the restriction against copying, and prevent this very common practice. There is concern that the combination of rising costs and wholesale copying may lead to the doom of the scientific journal, and possibly of some types of scientific books, in their present form. Numerous schemes have been suggested for preventing this situation or as alternatives to the present journal system, but none has yet been effective.

REFERENCES

1. Robert Calvert, "Patents (Practice and Management)," pp. 552–583 in Anthony Standen, ed., *Kirk–Othmer Encyclopedia of Chemical Technology*, 2nd ed., vol. 14, Wiley, New York, 1967.
2. U.S. Dept. of Commerce, Patent Office, *Patent Laws*, U.S. Government Printing Office, Washington, D.C., 1965.*
3. J. H. Saunders, *Careers in Industrial Research and Development*, Dekker, New York, 1974, Chap. 9.
4. Albert S. Davis, Jr., "A Piece of the Action," *Sci. & Tech.* (24), 49–53 (Dec. 1963).
5. Robert J. Kuntz, "The Proposed New Patent Law—Pro," *Res. Manage.* **15** (1), 64–68 (1972); C. Marshall Dann, "The Proposed New Patent Law—Con," *ibid.*, 68–71; Willard Marcy, "Rewarding Inventors in Academia," *ibid.*, 71–74.
6. Anon., "Patent Compensation Practices Vary Widely," *Chem. & Eng. News* **50** (39), 18–19 (Sept. 25, 1972).
7. U.S. Dept. of Commerce, Patent Office, *General Information Concerning Patents*, Stock No. 0304-0498, U.S. Government Printing Office, Washington, D.C., 1972.*
8. Arthur P. Kent, "Current Patent Office Procedures," *Chemtech* **2**, 599–605 (1972).
9. Günter Cramer, "The European Patent," *Chemtech* **4**, 670–674 (1974).
10. U.S. Dept. of Commerce, Patent Office, *Patents and Inventions: An Information Aid for Inventors*, Stock No. 0304-0511, U.S. Government Printing Office, Washington, D.C., 1972.*

* Available from the Superintendent of Documents, U.S. Government Printing Office, Washington, D.C. 20402.

11. U.S. Dept. of Commerce, Patent Office, *The Story of the U.S. Patent Office,* Stock No. 0304-0493, U.S. Government Printing Office, Washington, DC., 1972.*

12. Stacy V. Jones, *The Inventors Patent Handbook,* Dial, New York, 1969.

13. Terrence W. Fenner and James L. Everett, *Inventor's Handbook,* Chemical Publishing, New York, 1969.

14. Elmer J. Lawson and Edmund A. Godula, eds., *Patents for Chemical Inventions,* vol. 46 in Advances in Chemistry Series, Robert F. Gould, series ed., American Chemical Society, Washington, D.C., 1964.

15. Arthur J. Plantamura, "The Role of Patents in the Chemical Industry," Chap. 6 in Conrad Berenson, ed., *Administration of the Chemical Enterprise,* Wiley-Interscience, New York, 1963.

16. Errett S. Turner, "Patent Literature," pp. 583–635 in ref. 1.

17. U.S. Dept. of Commerce, Patent Office, *Official Gazette* (Patents Section), Weekly.*

18. U.S. Dept. of Commerce, Patent Office, *General Information Concerning Trademarks,* U.S. Government Printing Office, Washington, D.C., 1970.*

19. U.S. Dept. of Commerce, Patent Office, *Trademark Rules of Practice,* 8th ed., revision 1, U.S. Government Printing Office, Washington, D.C., 1973.*

20. Library of Congress, *Copyright Law of the United States of America,* U.S. Government Printing Office, Washington, D.C., 1972.*

21. Nicholas L. Henry, "Copyright: Its Application in Technological Societies," *Science* **186**, 993–1004 (1974).

* Available from the Superintendent of Documents, U.S. Government Printing Office, Washington, D.C. 20402.

CHAPTER 11

MANAGEMENT

The operating and staff divisions of a company, whose functions are discussed in Chaps. 6–9, must be integrated into an effective whole if the goals of the corporation are to be met. This integration and the direction of the company are provided by its management, and it represents a difficult and challenging assignment.

But at this stage you, the young professional entering industry, are far removed from a management career, and you may rightly wonder about the value of even a brief discussion of industrial management to you. Here are some of our reasons for including this chapter:

- Management establishes the environment in which you will be working in industry. Management's attitude can either inspire or stifle your creative and innovative talents. Your job can be a joy or drudgery depending on the company's management policies.
- If times get tough in future years, the question of whether you will retain your job or fall victim to a layoff will depend on management's policies. A company with a past record of layoffs of newly hired professionals in bad times may retain that policy in the future.
- Whether the company itself survives the challenges ahead will depend greatly on how rapidly and in what ways its management can adapt to change. A company with stagnant and unresponsive management will not prosper long.
- Management overly preoccupied with profits may make unethical compromises of its responsibilities to society—and may expect its employees to compromise their ethics also.

Thus we feel it is important for you to develop sufficient understanding of the functions, responsibilities, and problems of management that you can properly evaluate company management as an essential part of the industry you are entering.

In writing about management practice, one author comments (1) that "despite its crucial importance, its high visibility and its spectacular rise, management is the least known and the least understood of our basic institutions. . . . Even the people in a business often do not know what their management does and what it is supposed to be doing, how it acts and why, whether it does a good job or not." This is often not far from the truth, and you will recognize that we can't do more here than skim the surface of the topic of management. We will treat only the areas we think are most important to your understanding of how industry works. We will try to emphasize those factors that will aid you in assessing the management of the company you are interested in. We have already discussed how you can ultimately advance into management, in Chap. 5.

THE EVOLVING JOBS OF MANAGEMENT AND MANAGERS

As is the case with many other terms we have come across in our study of the structure of industrial concerns, management does not have a simple, broadly accepted definition. Most people associate management and managers with company leadership and "the boss"; obviously the terms imply the direction and coordination of the activities of others to achieve the goals associated with leadership. But what are these goals, and what are the inputs to the management process which gives it a basis for action? We like to think of them in terms of the concepts described in Fig. 45.

Within the corporate organization there are many levels of management with varying degrees of authority. As a young professional, you

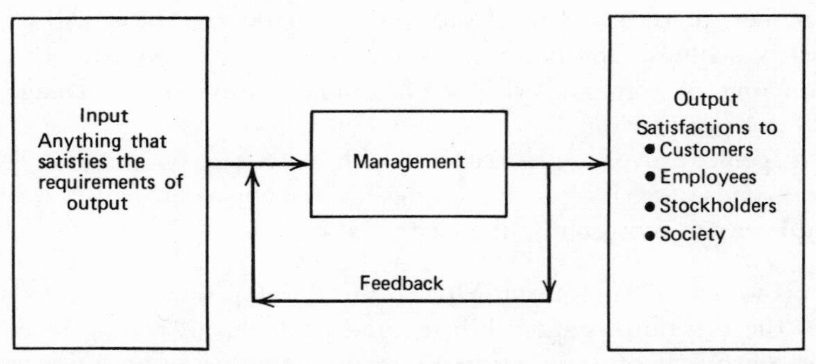

FIGURE 45. Management's job [adapted from (2)]. (Reprinted by permission of author and the American Chemical Society.)

may at once assume management responsibilities in directing the work of technicians or support personnel. As you advance within the organization from scientist or engineer to group leader or supervisor, section manager, division director, and on up the ladder, you will find that the management job itself remains essentially unchanged; the changes that occur with advancement are associated with the inputs and outputs (2).

There is no question that the principles of management have evolved greatly over the last few decades. There are at least six schools or types of management currently practiced, which we shall only mention by name; ref. 3 provides detail. They are the operational (management process), empirical or case, human behavior, social system, decision theory, and mathematical schools. The recent advances in the science of management have provided a better understanding of the requirements for effective management, and all indications are that these advances will continue in the years to come. Management has been, is, and will for some time be a very dynamic field associated with changing concepts and new tools and resources, even though the basic jobs to be done haven't changed.

The jobs still involve the inputs—the facts on which action must be based—and the outputs—the answers resulting from the actions. As Irving S. Shapiro, chairman of the board for Du Pont, said recently (4), "A good manager must be able to sort out facts in such a way that he knows which are crucial to an issue. The answers aren't that difficult once you get the facts. The real trick is to know what your goals are and know how you can reach them.

"Of course, the other part of this is that you've got to work your tail off. . . ."

Perhaps it is this last requirement that makes some managers wish that the device shown in Fig. 46 were a standard tool to aid management decisions.

THE FUNCTIONS OF MANAGEMENT

Managerial functions fall roughly into the five categories of planning, organizing, staffing, directing, and controlling.

PLANNING

The planning function of management is basic to the wellbeing of the corporation, for without a plan and well-established goals, the company could flounder. Planning involves establishing goals and objectives and

Now that you are retiring, J. B., the board of directors wants to know if that's your crystal ball or the company's.

FIGURE 46. From (5). Cartoon by Henry R. Martin. By permission of the author.

the associated policies, strategies, and procedures for meeting these goals at all levels of the corporation. It requires decisions about the future aims of the company, after due consideration of numerous alternatives.

Planning is no less important in industry than it is in our daily lives. Here are some major rules for planning (3), most of which apply to both extremes:

- Planning must not be left to chance. Without proper implementation and the necessary facilities for it, plans fail.

- Planning must start at the top. "Certainly, no subordinate should be unduly critical of his superior without having come up with his [own] program, recommended it, pressed for it, and been able to defend it" (3).
- Planning must be organized. Planning and doing cannot be separated. Doing is directly guided by associated company policies which have been established by management. Many of these policies and practices can affect you directly; it is important that you become aware of them as soon as possible. An example is the interplay of policies affecting decisions in new-product development depicted in Fig. 47.

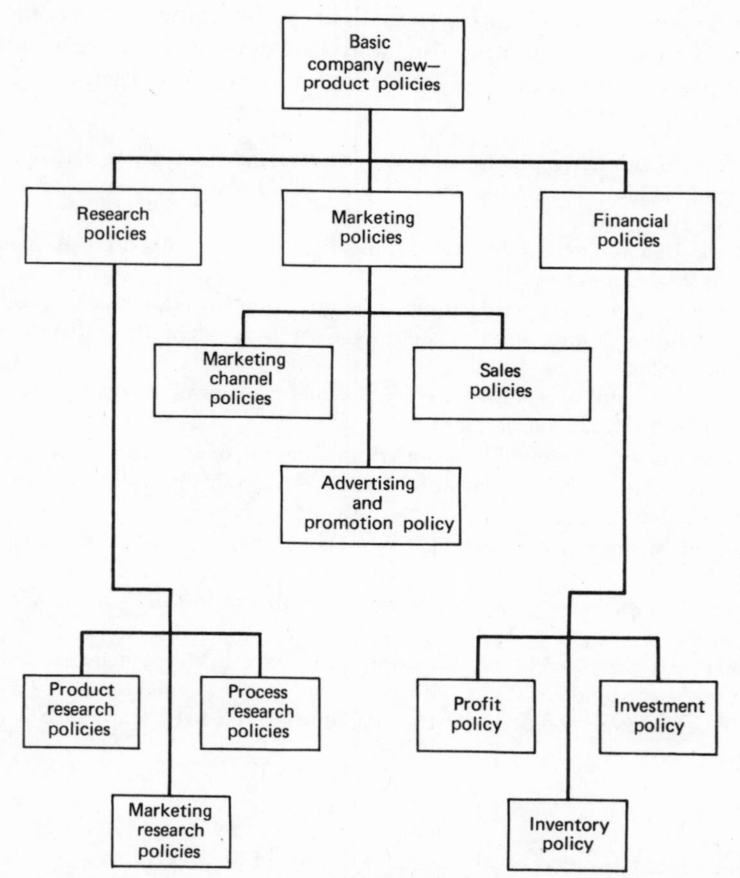

FIGURE 47. Hierarchy of policy in new-product development. (From *Principles of Management: An Analysis of Managerial Functions*, 4th ed., by Harold Koontz and Cyril O'Donnell. Copyright © 1968 by McGraw-Hill. Used with the permission of the McGraw-Hill Book Company.)

- Planning must be definite. The more specific plans are, the better they can be carried out, and the easier it is to determine what needs to be provided and to be done.
- Goals, premises, and policies must be communicated. The better the communication to and understanding by those who are expected to carry the plan out, the greater its chances of success.
- Long- and short-range plans must be integrated.
- Planning must include awareness and acceptance of change. Especially in these years of drastic economic upheaval, the importance of the company's ability to adapt to change must not be overlooked.

Change itself, now more than ever a corporate way of life, must not be overlooked as an essential ingredient of planning. But to the employee, change is often upsetting and threatening. Both management and employees would do well to bear in mind these guidelines (6) concerning change*:

> Change is more acceptable when it is understood than when it is not.
>
> Change is more acceptable when it does not threaten security than when it does.
>
> Change is more acceptable when those affected have helped create it than when it has been externally imposed.
>
> Change is more acceptable when it results from an application of previously established impersonal principles than it is when it is dictated by personal order.
>
> Change is more acceptable when it follows a series of successful changes than when it follows a series of failures.
>
> Change is more acceptable when it is inaugurated after prior change has been assimilated than when it is inaugurated during the confusion of another major change.
>
> Change is more acceptable if it has been planned than if it is experimental.
>
> Change is more acceptable to people new on the job than to people old on the job.
>
> Change is more acceptable to people who share in the benefits of change than to those who do not.
>
> Change is more acceptable if the organization has been trained to accept change.

ORGANIZING

Once a plan has been established, the people and activities required to meet its objectives must be organized. Lines of authority and responsi-

* Used with the special permission of DUN'S REVIEW, April 1957. Copyright © 1957 by Dun & Bradstreet Publications Corporation.

bility and a communication network must be established. Similar activities must be grouped together for efficiency's sake, and the established organization of the company's operating and staff divisions provides the framework for these management activities.

We have seen in Chaps. 6–9 how these operating and staff divisions are organized and managed, and now it is time to examine one final organization chart, that for management itself. As Fig. 48 indicates, the chain of command is simple, beginning with the ultimate owners of the company—the stockholders, if it is a public corporation—who elect the board of directors, to whom the rest of the organization is responsible.

STAFFING

Management's staffing activity involves providing and maintaining the proper personnel to implement the corporate plans, consistent with the limits set by financial considerations and personnel availability, and with the organizational lines defined through the organization charts. This activity includes providing the right candidates for the positions,

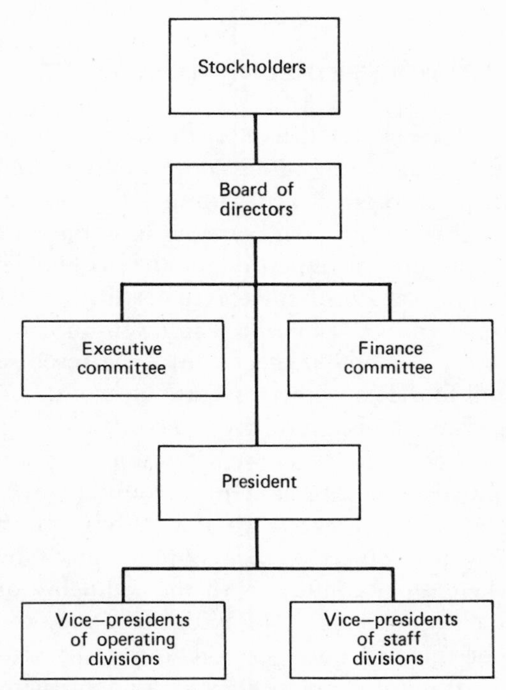

FIGURE 48. An organization chart for management.

specifying the requirements for the tasks (job descriptions), and providing for any specialized training required to develop the capabilities of the personnel to accomplish their specific tasks.

DIRECTING

Directing is simply guiding and supervising others in their pursuit of specified goals. The key ingredient in successful direction is not having authority, but using it properly. Among other things, authority must be used to motivate people, and any behavioral psychologist will attest to the difficulty of doing this. Means of applying authority to the direction of employees, and the relation to the results obtained, are part of the subject of management theory, treated briefly in a following section.

CONTROLLING

Controlling involves monitoring progress toward the goals set by the company's plans. Control must be exercised to keep costs from getting out of hand and to ensure that progress is not hampered by poor performance. Control involves overseeing and coordinating by management.

RESPONSIBILITIES OF MANAGEMENT

We devoted an entire chapter (Chap. 4) to the responsibilities you face on entering industry as a young professional. Similarly, management has large responsibilities associated with running the company. As Fig. 48 shows, the president is not his own boss, but is accountable to the board of directors, which, in turn, is responsible to the stockholders and through them to society. The nature of these responsibilities is similar to that outlined in Chap. 4, and we see no need to dwell on it here. You should remember that both you and management have responsibilities to the public interest and to the consumers, stockholders and employees.

Because these areas of responsibility sometimes overlap and at other times even conflict, there are some tough decisions about what should be done in many practical situations. One account (7) describes a stockholder's meeting at which an executive of a leading company stated that the traditional business objective of maximum profit had moved into "second place whenever it conflicts with the wellbeing of society." One stockholder jumped up to shout "Not with my money you don't!"

The reporter of this anecdote goes on to say (7) "Business can best begin discharging its social responsibility by humanizing its management practices. The company that acts on this insight and begins to operate on participatory forms of management is more likely to remain highly competitive." Coping adequately with the above responsibilities represents one of the greatest challenges for modern industrial management.

TRADITIONAL VS. HUMANISTIC MANAGEMENT

Unfortunately, managers are human and humans are fallible. A contributor to the book *The Failure of Success* (8) states that many of the problems of today's society result from incompetent organization and management (9)*:

> We have designed organizations which have ignored individual potential for competence, responsibility, constructive intent, and productivity. We have created structures and jobs that at the lower levels alienate and frustrate the workers; lead them to reject responsible behavior, not only with impunity but with a sense of justice; and tempt them to fight the organization by lowering the quality of what they produce. Products and services thus become tangible expressions of the low quality of life within organizations.
>
> Problems are equally severe at management levels, where incompetent organization structures create executive environments lacking in trust, openness, and risk taking. The attitudes that flourish best in such environments are conformity and defensiveness, which often find expression in an organizational tendency to produce detailed information for unimportant problems and invalid information for important ones.

Unfortunately, this is not undue pessimism but a true picture of the conditions that exist, to some extent, in many companies. In fact, the entirety of ref. 8 is devoted to possible ways that management can begin to correct the problems of organizational incompetence that plague our society today.

How the present state of affairs, involving so much incompetence in management, came about and what can be done about it are very difficult questions to answer. They involve the philosophy of how to manage, and we now examine this philosophy briefly to elucidate these problems. A partial answer, at least, to the first question comes from examining management philosophy as it existed for the greater part of this century.

* Reprinted by permission of the publisher from Alfred J. Morrow, ed., *The Failure of Success*. Quote by Chris Argyris (9). Copyright © 1972 by AMACOM, a division of American Management Associations.

TRADITIONAL MANAGEMENIT: THEORY X

McGregor (10) has characterized the traditional view of management, which he calls Theory X, in terms of the assumptions about human behavior which guide management's actions in directing people. His assumptions for Theory X are*:

1. The average human being has an inherent dislike of work and will avoid it if he can.
2. Because of this human characteristic of dislike of work, most people must be coerced, controlled, directed, threatened with punishment to get them to put forth adequate effort toward the achievement of organizational objectives.
3. The average human being prefers to be directed, wishes to avoid responsibility, has relatively little ambition, wants security above all.

This is certainly an uncomplimentary picture, but it does provide an explanation of some of the behavioral patterns one sometimes sees in industry. Of course, it is difficult to establish whether Theory X is correct because of the natural tendency of people to behave this way, or whether people are led to behave this way as a consequence of management based on Theory X (11). McGregor ascribes to the second point of view.

The ancient "carrot and stick" approach to motivate (Fig. 49) is an example of Theory X management.

HUMANISTIC MANAGEMENT: THEORY Y

McGregor's alternative to traditional management is a humanistic approach which he calls Theory Y. Here the guiding assumptions are (10)*:

1. The expenditure of physical and mental effort in work is as natural as play or rest.
2. External control and the threat of punishment are not the only means for bringing about effort toward organizational objectives. Man will exercise self-direction and self-control in the service of objectives to which he is committed.
3. Commitment to objectives is a function of the rewards associated with their achievement.
4. The average human being learns, under proper conditions, not only to accept but to seek responsibility.
5. The capacity to exercise a relatively high degree of imagination, ingenuity, and creativity in the solution of organizational problems is widely, not narrowly, distributed in the population.

FIGURE 49. From (5). Cartoon by Henry R. Martin. By permission of the author.

6. Under the conditions of modern industrial life, the intellectual potentialities of the average human being are only partially utilized.

The assumptions of Theories X and Y, amplified and paraphrased, are contrasted in Table 25.

The managerial strategy of Theory Y recognizes the possibility of human growth and development and stresses the need for selective adaptation rather than absolute control. It is a dynamic rather than a static strategy, pointing out that "the limits on human collaboration in the

TABLE 25

Human Behavior Assumptions of Theories X and Y*

Theory X	Theory Y
People by nature:	People by nature:
1. Lack integrity	1. Have integrity
2. Are fundamentally lazy and desire to work as little as possible	2. Work hard toward objectives to which they are committed
3. Avoid responsibility	3. Assume responsibility within these commitments
4. Are not interested in achievement	4. Desire to achieve
5. Are incapable of directing their own behavior	5. Are capable of directing their own behavior
6. Are indifferent to organizational needs	6. Want their organization to succeed
7. Prefer to be directed by others	7. Are not passive and submissive
8. Avoid making decisions whenever possible	8. Will make decisions within their commitments
9. Are not very bright	9. Are not stupid

* Adapted originally from *The Human Side of Enterprise* by Douglas McGregor. Copyright © 1960 by McGraw-Hill. Used with permission of the McGraw-Hill Book Company.

organizational setting are not limits of human nature but of management's ingenuity in discovering how to realize the potential represented by its human resources. Theory X offers management an easy rationalization for ineffective organizational performance: It is due to the nature of the human resources with which we must work. Theory Y, on the other hand, places the problems squarely in the lap of management . . . [implying] that the causes lie in management's methods of organization and control" (10).

Theory Y opens up an entire new range of management policies and practices, some of which may never be achieved. If management remains bound to Theory X, it may never recognize the extent of the potentialities of the human resources with which it works; if it accepts the assumptions of Theory Y, it will be challenged to discover and apply new ways of organizing and directing human effort.

HUMAN NEEDS AND MOTIVATION

We have commented several times on the importance of motivation in ensuring employee satisfaction and productivity. Since motivation is a management problem—many would say a management responsibility—we turn again to the subject.

Current thinking on motivation recognizes two major concepts (12): The first is concerned with influences that are considered motivators, such as recognition, achievement, job satisfaction, and advancement. When a manager knows how to create an atmosphere in which people can find these satisfactions, they will be motivated to respond favorably and strive for higher levels of accomplishment. The second concept relates the type of management required to get the best out of a person to his level of personal development, both intellectual and emotional. At the lowest and simplest levels, the individual needs clear, authoritative direction; as he matures, he needs to participate more in decision making and to express his individuality.

Man is motivated by his own needs as he sees them, and these fall into different levels, from the most basic to the most sophisticated (12):

- Physiological needs—food, rest, exercise, shelter.
- Safety needs—protection against danger, threat, deprivation.
- Social needs—association, acceptance, belonging, friendship, love.
- Ego needs—self-confidence, independence, achievement, knowledge, status, recognition, appreciation, respect.
- Self-fulfillment needs—realization of potentialities, self-development experience, creativity.

Obviously, most young professionals will have advanced to the point where their motivators are the needs associated with the higher levels of ego and self-fulfillment.

As management faces the problems associated with motivation—not the least of which is the growing alienation of the worker, illustrated by lower productivity, poorer workmanship, greater absenteeism, more turnover—its emphasis must be on providing greater job satisfaction by restructuring the organization to more closely meet the needs of its employees at all levels. Some of the new techniques that are beginning to be used are (13):

- Job rotation, in which the employee is moved through a series of related assignments.
- Job enlargement, in which tasks normally done by several employees are combined into a sequence performed by one worker.
- Job enrichment, in which the employee assumes some responsibility for planning and controlling his job as its level of difficulty or complexity is increased.

- Job styling, in which the job content is modified to provide the best fit of the employee to his assignment.
- Job-man measurement, in which the needs of individuals and the potentials of jobs to satisfy them are measured and matched.
- Task progression, in which job performance levels are precisely defined and sequenced.
- Team building, which seeks to increase a group's motivation to do its job well.

References 10, 11, and 14–17, written by the recognized leaders of human motivation analysis today, are recommended for further study.

MANAGEMENT TECHNIQUES

Many other management techniques have become popular in recent years, but to discuss them is beyond the scope of this book. Among them are management by objectives (18, 19), sensitivity training (20), and participative management (21).

WHAT TO LOOK FOR IN COMPANY MANAGEMENT

In this section we make some suggestions to help you in assessing the management characteristics of the company in which you are interested. If you are presently a candidate for a job, this review of company management policies should be a part of the more comprehensive assessment of companies described in Chap. 3.

MANAGEMENT PATTERNS

Companies and industries develop and adhere to patterns of management operation, which go a long way to establishing their reputations in many cases. If the pattern is well established, it will probably continue rather than undergo abrupt change.

- Look out for companies with past records of frequent layoffs of professionals. *Chemical & Engineering News* periodically reports the circumstances surrounding layoffs of chemists and chemical engineers.
- Watch out for companies with a history of labor unrest and labor–management difficulties.
- Seek out companies with a reputation for progressive management, as described below.

- Be aware of industries with high pay but low job security, such as aerospace, which are often those highly dependent on government funding.

PROGRESSIVE MANAGEMENT

Here are some suggestions on how to look for progressive management:

- Examine the corporate annual report. Is the company in deep financial trouble? Do the programs under way reflect humanistic management? Does management make the necessary bold decisions associated with new high-risk business projects (Chap. 6) and declining products (Chap. 8)? Are company profits rising or falling with respect to those of the industry as a whole, and why? Companies have growth curves, just as do products; in what stage of its growth curve is the company which you are evaluating?
- Look up the company's financial data in *Moody's Industrial Manual*, (22), available in most libraries.
- Read current and annual review articles in such industry news and trade magazines as those cited in Chap. 3 and in such business and management journals as *Business Week* (23), *Harvard Business Review* (24), and *Dun's Review* (25).

HUMANISTIC MANAGEMENT

Consider the following questions to determine just how humanistic top management's attitudes and practices are:

- Do the employees you have talked to give the impression that they are happy and stimulated by their jobs?
- Are they enthusiastic about their jobs, company management, and the company's future?
- Do they participate in management decisions?
- Are management attitudes closer to those of Theory X or Theory Y?
- How is the company adapting to change?
- What about job enrichment and related techniques?
- Does the company encourage its employees to be active in such outside activities as community projects?
- What about EEO? Does the company have an affirmative action program?
- What is the company doing to help its employees and the community?

ORGANIZATIONAL MALFUNCTION

Look for signs of organizational malfunction (2):

- Strength of the leaders. In a good company, the key executives stand out as leaders, having the respect of their employees and coordinating their department's work in a progressive manner. In a bad one, there is chaos instead of directed management, and the top executives spend their time trying to restore harmony.
- Too many levels of management. In a good company the path of authority flow and responsibility is short and uncluttered. The required number of levels varies, but Sears, Roebuck operates with just two levels between a section manager in a retail store and the president!
- Too many coordinators. The need for a "coordinating force" of liaison men, staff assistants, committees, and expediters is a sure sign of organizational malfunction.
- Going through channels. If you want to know what was in the sample you submitted for analysis yesterday, can you telephone the lab and talk to the analyst, or must your supervisor get the result from his supervisor? If the area engineer sees a leaking valve in a sulfuric acid line, can he fix it at once by tightening the packing nut, or must he go through channels to avoid having a union strike or grievance on his hands? In other words, is the structure of the system the end, or the means to achieving company objectives?
- Unbalanced management age structure. There should be a good mixture of young, middle-aged, and older executives, so that promotion and succession can take place regularly and smoothly, and the experience of age can be integrated with the drive of youth.
- Too much functional delegation. Top management should not delegate too much authority to functional specialists. This in effect gives the professional too many bosses, too many people he has to satisfy.
- Failure to clarify authority relationships. Levels of responsibility and authority must go hand in hand; an executive should never make a man responsible for an assignment but not give him the authority to do the job.

SOME EXAMPLES OF WELL-MANAGED COMPANIES

In its December issue each year, *Dun's Review* (25) lists and awards "The Five Best-Managed Companies" in the United States. In 1972 (26), these were Du Pont, Kodak, Mobil, Pfizer, and Xerox; in 1973 (27), Citibank, Exxon, Monsanto, J. C. Penney, and Weyerhaeuser; in 1974 (28), AT&T,

Kerr-McGee, Merck, R. J. Reynolds, and Southern Railway. It is gratifying to see so many of these within the broader definitions of the chemical industry. To illustrate the basis on which these awards are made, we cite an example:

Monsanto Company (29)

Like the other 1973 winners of Dun's awards, Monsanto recognized a serious management problem a few years ago and took determined action to overcome it. At Monsanto it was recognition of an unwieldy corporate structure that led to reorganization in 1971 and a new president, John Hanley, taking over in November 1972. As a result, Monsanto made one of the most startling comebacks in all industry, almost doubling its earnings ($6.90 per share in 1973 vs. $3.49 in 1972) in comparison with an average 32% gain for five of its top chemical competitors.

Jack Hanley had to be brought into the company from outside, and one of his top priorities stemmed from the fact that there was no strong in-company candidate for the job that became his. He made a number of moves toward more flexible and responsive management: moving decision-making responsibilities further down the line, setting up a new office of organization and management development, and instituting a system in which key executives outline their own goals for the year.

Although Monsanto has formidable resources in basic chemicals, Hanley is moving it further toward goods and chemical end products, where "value added" and profits are greater. New yarns for panty hose, LED's for digits on pocket calculators, plastic soft-drink bottles, and herbicides are some of the new directions for the company; old and declining products such as saccharine (a shocker, since it was Monsanto's first product) and polyester tire cord have been abandoned, and the facilities used to make L-Dopa and nylon carpet yarn, respectively, instead. And with all this, the company's marketing, administrative and technological expenses were reduced from 16.6% of sales in 1969 to 13% in 1973.

CHALLENGES AHEAD

As in Chaps. 6–8, we close this discussion with a look at some of the problems that must be faced in the near future, in this case by industrial management. Many of the problems faced by management are the same ones we've met before, and may appear in different guises as various authors write about them.

A recent issue of Du Pont's public affairs magazine *Context* (30) was devoted to management. In it, business writer John Thackray (31) identified the old familiar problems of the environment–pollution issue, the energy crisis, multinational operations, government regulations, and the social responsibilities of industry, as among the increasingly diverse challenges that management must face. Interviews with several business leaders (32) added to these the needs for a shift from quantity to quality in products and services; for careful assessment of the effects of introducing, extending, and modifying technology on humans and the environment; for emphasis on performance and quality instead of just products; for a global outlook to take advantage of worldwide marketing and investment opportunities; for a transition to a better educated, younger, and more mobile work force; and for a changing style of management that will include both specialists and generalists. And you will enjoy a satirical article (33) contrasting today's mild-mannered (?) managers with the tyrants of past years.

Management educator Peter Drucker points out (34) that in this day of increasingly complex and diverse businesses, the old organizational structures no longer suffice. The classical functional organization of a small manufacturing business, which has formed the basis of much of our discussion in this book, is still valid—but there aren't nearly as many small classical manufacturing businesses as there were. Likewise, the large, multidivisional company organization exemplified by General Motors' structure still works well for big single-product, single-market corporations, but more and more big companies fail to fit this mold. Thus, many companies are, and others will be, in the throes of reorganization until new patterns more suited to today's and tomorrow's needs are developed.

Harry Levinson, noted for his application of clinical psychology concepts to such problems as executive stress, provides (35) a list of problems that worry executives: social leadership, dealing with young people, scientific management (which usually turns out to be upgraded industrial engineering), obsolescence, dependency on others, ambivalence (between the wish to act autonomously and the need to share power), prediction of future trends, and motivation.

Meanwhile, middle management is in revolt (36, 37). Middle managers feel they are given no real role in decision making, but are held responsible for goals that aren't accomplished—the result of job responsibility without real authority. They find that traditional management controls and authoritarian methods (Theory X) are still in full swing, and are fearful of being trapped in the same unrewarding job until retirement. Many of them are becoming more concerned with the quality of their lives than with career success, with their personal goals and their families instead of job security. Turning down promotions and transfers that would mean family disruptions and personal discomfort, or refusing to take actions that would compromise their ethics, they are quitting—and finding it easy to get new jobs with companies more responsive to their needs.

The problems of women in management remain severe (38, 39). Why is it that although women make up 40% of the U.S. labor force, only 8% of the managers and administrators are women? One postulate is that most professional women are so devoted to specialization, not only as a skill but a psychological investment, that they have great need for breadth and managerial training, yet fear the risks involved in giving up their special skills to learn a new job, even at a higher level. Without doubt, the overcoming of these attitudes is a real challenge to management of the future.

REFERENCES

1. Peter F. Drucker, *The Practice of Management*, Harper & Row, New York, 1954.
2. Conrad Berenson, "Management in the Chemical Industry," Chap. 2 in Conrad Berenson, ed., *Administration of the Chemical Enterprise*, Wiley, New York, 1963.
3. Harold Koontz and Cyril O'Donnell, *Principles of Management: An Analysis of Managerial Functions*, 4th ed., McGraw-Hill, New York, 1968.
4. Anon., "Du Pont's Chairman: Irving S. Shapiro," *Du Pont Context* **3** (1), 4–6 (1974).
5. H. Martin, *All Those in Favor*, American Management Association, New York, 1965.
6. Ralph M. Besse, "Company Planning Must Be Planned!" *Dun's Rev.* **74** (4), 46 (1957.)
7. Alfred J. Marrow, "The Failure of Success," pp. 8–19 in ref. 8.
8. Alfred J. Marrow, ed., *The Failure of Success*, AMACOM, Div. American Management Association, New York, 1972.
9. Chris Argyris, "A Few Words in Advance," pp. 3–7 in ref. 8.
10. Douglas McGregor, *The Human Side of Enterprise*, McGraw-Hill, New York, 1960.
11. Saul W. Gellerman, *Motivation and Productivity*, American Management Association, New York, 1963.
12. Russell F. Moore, ed., *AMA Management Handbook*, American Management Association, New York, 1970.
13. Delmar L. Landen and Howard C. Carlson, "New Strategies for Motivating Employees," pp. 177–187 in ref. 8.
14. A. H. Maslow, *Motivation and Personality*, Harper & Row, New York, 1957.
15. Frederick Herzberg and B. B. Snyderman, *Motivation to Work*, 2nd ed., Wiley, New York, 1959.
16. Frederick Herzberg, *Work and the Nature of Man*, World Publications, Cleveland, 1966.
17. Chris Argyris, *Personality and Organization*, Harper & Row, New York, 1957.
18. Walter S. Wikstrom, *Managing by—and with—Objectives*, Personnel Policy Study No. 212, National Industrial Conference Board, New York, 1968.
19. David N. Beach and Walter R. Mahler, "Management by Objectives," pp. 231–240 in ref. 8.

20. Leland P. Bradford, "How Sensitivity Training Works," pp. 241–256 in ref. 8.
21. Alfred J. Marrow, "Participation: How It Works," pp. 83–89 in ref. 8; Alfred J. Marrow, "The Effect of Participation on Performance," pp. 90–102 in ref. 8; Alfred J. Marrow, Stanley E. Seashore, and David G. Bowers, "Managing Major Change," pp. 103–119 in ref. 8.
22. Robert P. Hanson, ed., *Moody's Industrial Manual,* Moody's Investors Service, New York, annually.
23. *Business Week,* McGraw-Hill, Inc., 1221 Ave. of the Americas, New York, New York 10020.
24. *Harvard Business Review,* Graduate School of Business Administration, Harvard University, Boston, Massachusetts 02163.
25. *Dun's Review,* Dun & Bradstreet Publications Corp., 666 Fifth Ave., New York, New York 10019.
26. Anon., "The Five Best-Managed Companies," *Dun's Rev.* **100** (6), 33 (Dec. 1972).
27. Anon., "The Five-Best-Managed Companies," *Dun's Rev.* **102** (6), 51 (Dec. 1973).
28. Anon., The "Five Best-Managed Companies," *Dun's Rev.* **104** (6), 43 (Dec. 1974).
29. Anon., "Monsanto: The Chemistry of a Comeback," *Dun's Rev.* **102** (6), 60–63 (Dec. 1973).
30. *Du Pont Context,* Public Affairs Dept., E. I. du Pont de Nemours and Co., D-8111, 1007 Market St., Wilmington, Del. 19898.
31. John Trackray, "Management's New Frontiers," *Du Pont Context* **3** (1), 1–2 (1974).
32. Craig Garner, "Peering into Tomorrow: What the Future Holds for Managers," *Du Pont Context* **3** (1), 25–27 (1974).
33. John McArthur, "Today's Managers are Pussycats Compared with the Tigers of Yesteryear," *Du Pont Context* **3** (1), 22–23 (1974).
34. Peter F. Drucker, "New Templates for Today's Organizations," *Harvard Bus. Rev.* **52** (1), 45–53 (Jan.–Feb. 1974).
35. Harry Levinson, "Problems That Worry Executives," pp. 67–80 in ref. 8.
36. Thomas J. Murray, "The Revolt of the Middle Managers—Phase Two," *Dun's Rev.* **102** (2), 32–34 (August 1973).
37. Emanuel Kay, *Crisis in Middle Management,* AMACOM, Div. of American Management Association, New York, 1974.
38. Anon., "Women in Management: What Needs to be Done?" *Du Pont Context* **3** (1), 13–16 (1974).
39. Edith M. Lynch, *The Executive Suite—Feminine Style,* AMACOM, Div. American Management Association, New York, 1973.

CHAPTER 12

CONCLUSION

We cannot, in good conscience, close this book without a few last words. In part, we suppose, it is our pedagogical training: As we said in the Preface, we have told you what we wanted to tell you, and now we must tell you what we have told you.

We hope we have been successful in telling you young professionals what industry is all about, using the American chemical industry as our example: how it works, how to approach it and land your first job, what responsibilities you will face, how you will advance, and what you will find in its many divisions including R&D, manufacturing, marketing, the staff divisions, and management.

But there is more we wish to say. Life, we feel, requires all of us to meet a continuing series of challenges, and we have spent some time in telling you about those that we expect you—and industry—will be facing in the years to come. How can they be met successfully? We believe that this will require a new era of cooperation among all parts of society: industry, the universities, government, and the public. For the problems to be faced are common to all and interlock them inseparably together. Our concern here is with the cooperation that affects you most as young professionals: the academia–industry interface. As we said in Chap. 1, it almost does not exist now, but we feel it must be developed—not as a barrier but as a gateway—if we are to achieve the cooperation that is required to enhance the productivity, both material and personal, that young professionals can and must bring into industry through their academic training. An example we have used before is the need in industry for young professionals trained in polymer science and engineering, in contrast to the lack of interest in providing this training in most universities. On the occasion of the award of the 1974 Nobel Prize in chemistry to the polymer chemist Paul J. Flory, this need and corresponding lack of interest were presented as a challenge to the universities by Paul

"What everyone seems to forget is that A.T.&T., when you come right down to it, is only people—like you and me."

FIGURE 50. From *The New Yorker* **50**(47), 27 (January 13, 1975). Drawing by Weber. Copyright © 1975 by The New Yorker Magazine, Inc.

Lindenmeyer of the National Science Foundation in a *Chemical & Engineering News* editorial (1).

In addition to challenges there is the matter of goals. We hope we have been successful in describing some of the goals of industry, and in pointing out that profit for profit's sake is less and less often considered one of them. (And we might add that education for the sake of education is a hollow goal, too.) We hope this book has been of some help to you in selecting and confirming your own personal goals as they relate to your life's career, and that we have been able to suggest ways that will make meeting them a little easier. We are sure you will find industry sympathetic to your goals and anxious to aid you in achieving them—after all, as Fig. 50 suggests, no matter who we are we all have common wants and needs. In the undertaking of your professional career designed to meet your goals, we wish you all success.

If one were inclined to pessimism, he could no doubt find much in this book to justify his attitude. Crises surround us—recession, inflation, depression, energy shortages, taxes, unrest, and many more. But we subscribe to a different and more optimistic point of view, that expressed by Ben Wattenberg in his book *The Real America* (2). We think America is basically sound and one of the best places in the world to live. We feel that industry has played a major part in making it so, and will continue to contribute to the wellbeing of this country and its people. And we know that it is the young professional scientists and engineers who are the industrialists of today and the managers of tomorrow. It is in your hands as much as those of anyone that our future lies. If we have helped even a few of you to find the place in this future where you and all of us will prosper, then we will consider our efforts successful.

REFERENCES

1. Paul H. Lindenmeyer, "Wake Up and Smell the Coffee," *Chem. & Eng. News* **52** (46), 4, 54 (Nov. 18, 1974).
2. Ben Wattenberg, *The Real America,* Doubleday, New York, 1974.

APPENDIX 1

GUIDELINES TO PROFESSIONAL EMPLOYMENT FOR ENGINEERS AND SCIENTISTS*

FOREWORD

This publication is a guide to mutually satisfying relationships between professional employees and their employers. In this document, professional employees are defined as engineers and scientists. These Guidelines cover factors peculiar to professional employment, and omit many generally accepted precepts of personnel relations which are common to all classifications of employees.

These Guidelines are applicable to professional employment in all fields and in all areas of practice (including both non-supervisory and supervisory positions), and are based on the combined experience and judgment of all of the endorsing societies.

It must be stressed in the implementation of these Guidelines that they represent desirable general goals rather than a set of specific minimum standards. Wide variations in circumstances and individual organizational practices make it inappropriate to judge any given employer on the basis of any single employment policy or fringe benefit. Rather, attention should be devoted to evaluating the entire employment "package," including such intangibles as opportunity for future advancement or participation in profits, location, local cost of living, and other factors which may be important to professional employees.

Observance of the spirit of these Guidelines will minimize personnel problems, reduce misunderstandings, and generate greater mutual respect. It is anticipated that they will be of use to employers in evaluating their own practices, to professional employees in evaluating both their

* Reprinted from *Guidelines to Professional Employment for Engineers and Scientists*, January 1, 1973. Copies available upon request from endorsing societies: AACE, AIChE, AIIE, ANS, ASCE, ASME, ECPD, EJC, IEEE, ITE, NICE, NSPE, and SFPE.

own responsibilities and those of their employers, and to new graduates and other employment seekers in obtaining a better picture of prospective employers. Where differences in interpretation occur, they may be referred to the headquarters office of any of the endorsing societies.

OBJECTIVES

The endorsing societies, with their avowed purpose to serve the public and their professions, recognize clearly that in order to make a maximum contribution, it is necessary for professional employees and employers to establish a climate conducive to the proper discharge of mutual responsibilities and obligations.

Essential and prerequisite to establishing such a climate are:

1. Mutual loyalty, cooperation, fair treatment, ethical practices, and respect are the basis for a sound relationship between the professional and his employer.
2. The professional employee must be loyal to the employer's objectives and contribute his creativity to those goals.
3. The responsibility of the professional employee to safeguard the public interest must be recognized and shared by the professional employee and employer alike.
4. The professional growth of the employee is his prime responsibility, but the employer undertakes to provide the proper climate to foster that growth.
5. Factors of age, race, religion, political affiliation, or sex should not enter into the employee/employer relationship.

Effective use of these Guidelines is accomplished when the employer provides each present and prospective professional employee with a written statement of his policies and practices relating to each of the items covered. Adherence to these guidelines by employers and professional employees will provide an environment of mutual trust and confidence. Local conditions may result in honest differences in interpretation of, and in deviation from, the details of these guidelines. Such differences should be resolved by discussions leading to an understanding which meets the spirit of the guidelines.

I. RECRUITMENT

Employment should be based solely on professional competence and ability to adequately perform assigned responsibilities, with employee qualifications and employment opportunities represented in a factual and forthright manner. The employer's offer of employment and the employee's acceptance, should be in writing, including a clear understanding with regard to relocation assistance; past, present and future confidentiality and patent obligations; salary; expected duration of employment; and other relevant employment conditions and benefits.

PROFESSIONAL EMPLOYEE

1. The professional employee (applicant) should attend interviews and accept reimbursement only for those job opportunities in which he has a sincere interest. The applicant should prorate costs for multiple interviews during a given trip on a rational basis. The guiding principle should be that the applicant receives neither more nor less than the cost of the total trip.

2. The applicant should carefully evaluate past, present, and future confidentiality obligations in regard to trade secrets and proprietary information connected with the potential employment. He should not seek or accept employment on the basis of using or divulging any trade secrets or proprietary information.

3. Having accepted an offer of employment, the applicant is morally obligated to honor his commitment unless formally released after giving adequate notice of intent.

4. The applicant should not use the funds or time of his current employer for the purpose of seeking new employment unless approved by the current employer.

EMPLOYER

1. The policy of the employer regarding payment of expenses incurred by the applicant in attending the interview must be made clear prior to the arranged interview.

2. The applicant should have an interview with his prospective supervisor in order to understand clearly the technical and business nature of the job opportunity. This prospective supervisor should be ethically responsible for all representations regarding the conditions of employment.

3. Applications for positions should be confidential. The expressed consent of the applicant should be obtained prior to communicating with a current employer.

4. Employers should minimize hiring during periods of major curtailment of personnel. Hiring of professional personnel should be planned at all times to provide satisfying careers.

5. Agreements among employers or between employer and professional employee which limit the opportunity of professional employees to seek other employment or establish independent enterprises are contrary to the spirit of these guidelines.

6. Having accepted an applicant, an employer who finds it necessary to rescind an offer of employment should make adequate reparation for any injury suffered.

II. TERMS OF EMPLOYMENT

Terms of employment should be in writing, in accordance with the applicable laws, and consistent with generally accepted ethical professional practices.

PROFESSIONAL EMPLOYEE

1. The professional employee should be loyal to his employer. He should accept only those assignments for which he is qualified; should diligently, competently, and honestly complete his assignments; and he should contribute creative, resourceful ideas to his employer while making a positive contribution toward establishing a stimulating work atmosphere and maintaining a safe working environment.

2. The professional employee should have due regard for the safety, life, and health of the public and fellow employees in all work for which he is responsible. Where the technical adequacy of a process or product is involved, he should protect the public and his employer by withholding approval of plans that do not meet accepted professional standards and by presenting clearly the consequences to be expected if his professional judgment is not followed.

3. The professional employee should be responsible for the full and proper utilization of his time in the interest of his employer and the proper care of the employer's facilities.

4. The professional employee should avoid any conflict of interest with his employer, and should immediately disclose any real or potential problem which may develop in this area. He should not engage in any other professional employment without his employer's permission.

5. The professional employee should not divulge technical proprietary information while he is employed. Furthermore, he should not divulge or use this information for an agreed upon period after employment is terminated.

6. The professional employee should only sign or seal plans or speci-

fications prepared by himself or others under his supervision, or plans or specifications that he has reviewed and checked to his personal satisfaction.

7. The professional employee should not accept payments or gifts of any significant value, directly or indirectly, from parties dealing with his client or employer.

EMPLOYER

1. The employer should inform his professional employees of the organization's objectives, policies and programs on a continuing basis.

2. The professional employee should receive a salary in keeping with his professional contribution which reflects his abilities, professional status, responsibility, the value of his education and experience, and the potential value of the work he will be expected to perform. The salary should be commensurate with the salaries of other employees both professional and nonprofessional. Sound indirect compensation programs should be provided. The most important are retirement plans, health and life insurance, sick leave, paid holidays and paid vacations.

3. The employer should establish a salary policy, taking into account published salary surveys, and provide equitable compensation for each employee commensurate with his position and performance. The salary structure should be reviewed annually to keep the assigned dollar values adjusted to the current economy.

4. Each individual position should be properly classified as to its level in the overall salary structure. The evaluation of each position should consider such factors as the skill required for acceptable performance, the original thinking required for solving the problems involved, and the accountability for an action and its consequences.

5. Economic advancement should be based upon a carefully designed performance review plan. Provisions should be made for accelerated promotions and extra compensation for special accomplishments. At least annually, performance evaluations and salary review should be conducted for the individual professional employee by his supervisor. Performance evaluations should include discussion on how well he has performed his work and what he can do to improve. The professional employee should be clearly informed if his performance is considered unsatisfactory. All promotions in salary and responsibility should be on an individual merit basis.

6. For the professional employee whose aptitude and interests are technical rather than supervisory, equivalent means of advancement and recognition should be provided.

7. It is inappropriate for a professional employee to use a time clock to record arrival and departure, particularly since situations may arise which require unusual effort on his part. However, if the work demanded of a professional employee regularly exceeds the normal working hours for extended periods, the employer should compensate him for his continuing extra effort according to a clearly stated policy.

8. The professional employee should be included in an adequate pension plan which provides for early vesting of rights in safeguarded pension funds. Vesting should be so scheduled that it does not seriously affect either the employer's or the professional employee's decision as to continued employment. As a goal, eligibility for participation should not exceed one year after employment, maximum full vesting time should be five years, and the minimum pension upon reaching retirement should be no less than 50% of the average best five years' salary (based on a forty-year working career with a single employer). If a pension plan is not provided, or the benefits are less than outlined above, other compensation should be increased proportionately.

9. The employer should provide office, support staff and physical facilities which promote the maximum personal efficiency of the professional employee.

10. Duties, levels of responsibility, and the relationship of positions within the organizational hierarchy should be clearly defined and should be accurately reflected in position titles.

11. The employer should not require the professional employee to accept responsibility for work not done under his supervision.

12. The employer should provide formal assurance through organizational policy that it will defend any suits or claims against individual professional employees employed by the organization in connection with their authorized professional activities on behalf of the employer.

13. There should be no employer policy which requires a professional employee to join a labor organization as a condition of continued employment.

14. It is the employer's responsibility to clearly identify proprietary information.

III. PROFESSIONAL DEVELOPMENT

The employee and the employer share responsibility for professional development of the employee—the employee to establish the goals and take the initiative to reach them, and the employer to provide the environment and attitude which is conducive to professional growth.

PROFESSIONAL EMPLOYEE

1. Each professional employee is responsible for maintaining his technical competence and developing himself through a program of continuing education.
2. The professional employee should belong to and participate in the activities of appropriate professional societies in order to expand his knowledge and experience. Such participation should include the preparation of professional and technical papers for publication and presentation.
3. The professional employee should achieve appropriate registration and/or certification as soon as he is eligible.
4. The professional employee should recognize his responsibility to serve the public by participating in civic and political activities of a technical and nontechnical nature. Such participation, however, should be undertaken solely as a responsibility of the individual without interfering with the timely execution of his work and without involving the employer.

EMPLOYER

1. The employer, as a matter of policy, should provide an atmosphere which promotes professional development. This will include, among other programs, encouraging and supporting membership and attendance at professional society meetings and at formal courses of study which will enable the employee to maintain his technical competence.
2. The employer should consider compensated leaves of absence for professional study as a means of enabling the employee to improve his competence and knowledge in a technical field.
3. Consistent with employer objectives, the employee should be given every opportunity to publish his work promptly in the technical literature and to present his findings at technical society meetings.
4. It is in the best interest of the employer to encourage continuing education to broaden the qualifications of employees through self-improvement, in-house programs, formal education systems in the institutions of higher learning, and meetings and seminars on appropriate subjects.
5. The employer should encourage and assist professional employees to achieve registration and/or certification in their respective fields.

IV. TERMINATION AND TRANSFER

Adequate notice of termination of employment should be given by the employee or employer as appropriate.

PROFESSIONAL EMPLOYEE

1. If the professional employee decides to terminate his employment, he should assist the employer in maintaining a continuity of function, and he should provide at least one month's notice. When termination is initiated by the employee, no severance pay is due.

EMPLOYER

1. Additional notice of termination, or compensation in lieu thereof, should be provided by the employer in consideration of responsibilities and length of service. As a desirable goal, permanent employees (after initial trial period) should receive notice or equivalent compensation equal to one month, plus one week per year of service. In the event that the employer elects notice in place of severance compensation, then he should allow the employee reasonable time and facilities to seek new employment.
2. Employers should make every effort to relocate terminated professional employees either within their own organizations or elsewhere. Consideration should be given to continuing major employee protection plans for some period following termination, and to their full reinstatement in the event of subsequent reemployment.
3. If a professional employee is involuntarily terminated on the basis of early retirement, the employer should consider an equitable provision for an adequate income for the period remaining until the employee receives his pension at his normal retirement age.
4. In a personal interview, the employer should inform the employee of the specific reasons for his termination.
5. The employer should provide an adequate transfer-time notice, with due consideration to the extent of personal matters which the professional employee must settle before moving. All normal costs of the transfer should be paid by the employer including moving expenses, realtor fees, travel expenses to the new location to search for housing, and reasonable living expenses for the family until permanent housing is found. Unusual moving expense reimbursement should be settled in a discussion between the employee and employer.

This document is subject to periodic review by the participating societies for the purpose of keeping it current. Suggested amendments will be considered collaboratively in connection with future revised editions.

APPENDIX 2

EMPLOYMENT AGREEMENT*

THIS AGREEMENT, entered into this date of 19 , between a corporation of (hereinafter called "Employer"), and (hereinafter called "Employee").

WITNESSETH:

WHEREAS, in its business, Employer has developed and uses commercially valuable technical and nontechnical information and, to guard the legitimate interests of Employer, it is necessary for Employer to protect certain of the information either by patents or by holding it secret or confidential; and

WHEREAS, the aforesaid information is vital to the success of Employer's business and Employee through his activities may become acquainted therewith, and may contribute thereto either through inventions, discoveries, improvements or otherwise;

NOW, THEREFORE, in consideration of and as part of the terms of employment of Employee by Employer, at a wage or salary and for such length of time as the Employment shall continue, it is agreed as follows:

1. Unless Employee shall first secure Employer's written consent, Employee shall not disclose or use at any time either during or subsequent to said employment, any secret or confidential information of Employer of which Employee becomes informed during said employment, whether or not developed by Employee, except as required in Employee's duties to Employer.

2. Employee shall disclose promptly to Employer or its nominee any and all inventions, discoveries and improvements conceived or made by Employee during the period of employment and related to the business

* From Studies in Personnel Policy No. 199. Reprinted by permission of The Conference Board.

or activities of Employer, and assigns and agrees to assign all his interest therein to Employer or its nominee; whenever requested to do so by Employer. Employee shall execute any and all applications, assignments or other instruments which Employer shall deem necessary to apply for and obtain Letters Patent of the United States or any foreign country or to protect otherwise Employer's interests therein. These obligations shall continue beyond the termination of employment with respect to inventions, discoveries and improvements conceived or made by Employee during the period of employment, and shall be binding upon Employee's assigns, executors, administrators and other legal representatives.

3. Upon termination of said employment, Employee shall promptly deliver to Employer all drawings, blueprints, manuals, letters, notes, notebooks, reports, and all other materials of a secret or confidential nature relating to Employer's business and which are in the possession or under the control of Employee.

IN WITNESS WHEREOF, the parties have signed this agreement in duplicate as of the date written above.

By ______________________________

______________________________ Dept.

____

EMPLOYEE

Witness:

____

APPENDIX 3

THE CHEMIST'S CREED*

AS A CHEMIST, I HAVE A RESPONSIBILITY:

TO THE PUBLIC

to propagate a true understanding of chemical science, avoiding premature, false, or exaggerated statements, to discourage enterprises or practices inimical to the public interest or welfare, and to share with other citizens a responsibility for the right and beneficent use of scientific discoveries.

TO MY SCIENCE

to search for its truths by use of scientific methods, and to enrich it by my own contributions for the good of humanity.

TO MY PROFESSION

to uphold its dignity as a foremost branch of learning and practice, to exchange ideas and information through its societies and publications, to give generous recognition of the work of others, and to refrain from undue advertising.

TO MY EMPLOYER

to serve him undividedly and zealously in mutual interests, guarding his concerns and dealing with them as I would my own.

TO MYSELF

to maintain my professional integrity as an individual, to strive to keep abreast of my profession, to hold the highest ideals of personal honor, and to live an active, well rounded, and useful life.

TO MY EMPLOYEES

to treat them as associates, being ever mindful of their physical and mental well-being, giving them encouragement in their work, as much freedom for personal development as is consistent with the proper

* Anon., "The Chemist's Creed," American Chemical Society, Washington, D.C., 1968. Reprinted with permission of the American Chemical Society.

conduct of work, and compensating them fairly, both financially and by acknowledgement of their scientific contributions.

TO MY STUDENTS AND ASSOCIATES

to be a fellow learner with them, to strive for clarity and directness of approach, to exhibit patience and encouragement, and to lose no opportunity for stimulating them to carry on the great tradition.

TO MY CLIENTS

to be a faithful and incorruptible agent, respecting confidence, advising honesty, and charging fairly.

APPENDIX 4

PROFESSIONAL SOCIETIES AND TRADE ASSOCIATIONS

American Association for the Advancement of Science, 1515 Massachusetts Ave., N.W., Washington, D.C. 20005

American Association of Textile Chemists and Colorists, Box 12215, Research Triangle Park, N.C. 27709

American Chemical Society, 1155 16th St., N.W., Washington, D.C. 20036

American Geological Institute, 2201 M St., N.W., Washington, D.C. 20037

American Institute of Aeronautics and Astronautics, 1290 Ave. of the Americas, New York, N.Y. 10019

American Institute of Biological Sciences, 3900 Wisconsin Ave., N.W., Washington, D.C. 20016

American Institute of Chemical Engineers, 345 E. 47th St., New York, N.Y. 10017

American Institute of Chemists, Inc., 7315 Wisconsin Ave., Washington, D.C. 20014

American Institute of Industrial Engineers, 345 E. 47th St., New York, N.Y. 10017

American Institute of Mining, Metallurgical, and Petroleum Engineers, 345 E. 47th St., New York, N.Y. 10017

American Institute of Physics, 335 E. 45th St., New York, N.Y. 10017

American Management Association, 135 W. 50th St., New York, N.Y. 10020

American Marketing Association, 230 N. Michigan Ave., Chicago, Ill. 60601

American Mathematical Society, Box 6248, Providence, R.I. 02904

American National Standards Institute, Inc., 1430 Broadway, New York, N.Y. 10018

American Nuclear Society, Inc., 244 E. Ogden Ave., Hinsdale, Ill. 60521

American Petroleum Institute, 1801 K St., N.W., Washington, D.C. 20036

American Pharmaceutical Association, 2215 Constitution Ave., N.W., Washington, D.C. 20037

American Physical Society, The, 335 E. 45th St., New York, N.Y. 10017

American Society of Civil Engineers, 345 E. 47th St., New York, N.Y. 10017

American Society for Engineering Education, One Dupont Circle, N.W., Suite 400, Washington, D.C. 20036

American Society of Mechanical Engineers, Inc., The, 345 E. 47th St., New York, N.Y. 10017

American Society for Metals, Metals Park, Ohio 44703

American Society for Quality Control, Inc., 161 W. Wisconsin Ave., Milwaukee, Wisc. 53202

American Society for Testing and Materials, 1916 Race St., Philadelphia, Pa. 19103

American Statistical Association, 806 Fifteenth St., N.W., Washington, D.C. 20005

Engineers Joint Council, Inc., 345 E. 47th St., New York, N.Y. 10017

Federation of American Societies for Experimental Biology, 9650 Rockville Pike, Bethesda, Md. 20014

Federation of Analytical Chemistry and Spectroscopy Societies, 4440 Warrensville Center Road, Cleveland, Ohio 44128

Federation of Societies for Coatings Technology, 121 S. Broad St., Philadelphia, Pa. 19107

Institute of Electrical and Electronic Engineers, Inc., 345 E. 47th St., New York, N.Y. 10017

Inter-Society Color Council, Rensselaer Polytechnic Institute, Troy, N.Y. 12181

Manufacturing Chemists Association, 1825 Connecticut Ave., N.W., Washington, D.C. 20009

Mathematical Association of America, 1225 Connecticut Ave., N.W., Washington, D.C. 20036

National Academy of Engineering, 2101 Constitution Ave., N.W., Washington, D.C. 20418

National Academy of Sciences, 2101 Constitution Ave., N.W., Washington, D.C. 20418

National Society of Professional Engineers, 2029 K St., N.W., Washington, D.C. 20006

Operations Research Society of America, 482 E. Preston St., Baltimore, Md. 21202

Society of Manufacturing Engineers, 20501 Ford Rd., Dearborn, Mich. 48128

Society of Plastics Engineers, 656 W. Putnam Ave., Greenwich, Conn. 06830

Society of the Sigma Xi, The, 345 Whitney Ave., New Haven, Conn. 06511

Society of Women Engineers, 345 E. 47th St., New York, N.Y. 10017

Synthetic Organic Chemical Manufacturers Association, 330 Madison Ave., New York, N.Y. 10017

Technical Association of the Pulp and Paper Industry, One Dunwoodie Park, Atlanta, Ga. 30341

INDEX